PENGUIN BOOKS

THE PENGUIN BOOK OF
CURIOUS AND INTERESTING PUZZLES

David Wells was born in 1940. He had the rare distinction of being a Cambridge scholar in mathematics and failing his degree. He subsequently trained as a teacher and, after working on computers and teaching machines, taught mathematics and science in a primary school and mathematics in secondary schools. He is still involved with education through writing and working with teachers.

While at university he became British under-21 chess champion, and in the middle seventies was a game inventor, devising 'Guerilla' and 'Checkpoint Danger', a puzzle composer, and the puzzle editor of *Games & Puzzles* magazine. From 1981 to 1983 he published *The Problem Solver*, a magazine of mathematical problems for secondary pupils.

He has published several books of problems and popular mathematics, including *Can You Solve These?* and *Hidden Connections, Double Meanings*, and also *Russia and England, and the Transformations of European Culture*. He has written *The Penguin Dictionary of Curious and Interesting Numbers* and *The Penguin Dictionary of Curious and Interesting Geometry*, and is currently writing a book on the nature, learning and teaching of mathematics.

David Wells

The Penguin Book of
Curious and Interesting Puzzles

PENGUIN BOOKS

PENGUIN BOOKS

Published by the Penguin Group
Penguin Books Ltd, 27 Wrights Lane, London W8 5TZ, England
Penguin Books USA Inc., 375 Hudson Street, New York, New York 10014, USA
Penguin Books Australia Ltd, Ringwood, Victoria, Australia
Penguin Books Canada Ltd, 10 Alcorn Avenue, Toronto, Ontario, Canada M4V 3B2
Penguin Books (NZ) Ltd, 182–190 Wairau Road, Auckland 10, New Zealand

Penguin Books Ltd, Registered Offices: Harmondsworth, Middlesex, England

First published 1992
10 9 8 7 6 5 4 3 2 1

The acknowledgements on p. vi constitute an extension of
this copyright page

The moral right of the author has been asserted

Typeset by DatIX International Limited, Bungay, Suffolk
Filmset in Monophoto Sabon
Printed in England by Clays Ltd, St Ives plc

Contents

Acknowledgements

Please note that detailed sources for puzzles are given at the end of each puzzle solution, where appropriate.

Grateful acknowledgement is given to the following:

Dover Publications, Inc., for permission to reproduce material from: Stephen Barr, *Second Miscellany of Puzzles* (1969); A. H. Beiler, *Recreations in the Theory of Numbers* (1966); Angela Dunn (ed.), *Mathematical Bafflers* (1980), and *The Second Book of Mathematical Bafflers* (1983); L. A. Graham, *Ingenious Mathematical Problems and Methods* (1959), and *The Surprise Attack in Mathematical Problems* (1968); J. A. H. Hunter, *More Fun with Figures* (1966); F. Mosteller, *Fifty Challenging Problems in Probability* (1987); F. Schuh, *The Master Book of Mathematical Recreations* (1968); George J. Summers, *New Puzzles in Logical Deduction* (1968).

I will also note here that although the original Loyd and Dudeney books are long out of print, two collections of Loyd's puzzles, both edited by Martin Gardner, are published by Dover under the titles *Mathematical Puzzles of Sam Loyd* and *More Mathematical Puzzles of Sam Loyd*, and they have also reprinted Dudeney's *Amusements in Mathematics*.

Robert Hale Ltd, for permission to reproduce 'Room for More Inside', from Gyles Brandreth, *The Complete Puzzler* (1982).

McGraw-Hill, Inc., for permission to reproduce 'The Truel', from David Silverman, *Your Move* (1971).

Weidenfeld and Nicholson for permission to reproduce the Tangram puzzles from E. Cuthwellis (ed.), *Lewis Carroll's Bedside Book* (1979).

John Hadley for the translation of Alcuin's *Propositiones ad acuendos juvenes*, and David Singmaster for lending me his copy, as well as giving me the run of his library of mathematical recreations. John Hadley's complete translation has subsequently been published in the *Mathematical Gazette*, Vol. 76, No. 475, March 1992.

Finally, I should like to thank the staff of the British Library for their courteous help.

Introduction

To puzzle and be puzzled are enjoyable experiences, so it is no surprise that puzzling problems are as old as history itself, and follow a similar pattern of many centuries of slow progress, followed by rapid expansion in the nineteenth century, and an explosion in the twentieth. This book follows that pattern. The first third is devoted to puzzles from the dawn of history, in Egypt and Babylon, up to the nineteenth century. These are followed by examples of the puzzles of Loyd and Dudeney, who straddle the nineteenth and twentieth centuries, and other famous puzzlers of that era such as Lewis Carroll and Eduard Lucas. The second half of the book is devoted to the great variety of puzzles composed in the twentieth century.

I must emphasize, however, that this is not a history. I have merely selected some representative figures. One day a history of puzzles will be written, I hope by David Singmaster, who has spent many years delving into the origins of popular puzzles, but in the meantime this book will give readers examples, only, of the puzzling questions that have found popular favour over the centuries.

Limitations of space have forced a strict selection. Word puzzles are entirely excluded. I hope that in due course they will form a separate volume, well justified by their immense richness and variety. A boundary also had to be drawn between puzzles of a logical and mathematical nature, and mathematical recreations and mathematics itself. This boundary cannot be drawn precisely, but generally speaking problems which require any mathematics beyond the most elementary algebra and geometry, have been excluded, and few of the puzzles require even that level of sophistication.

A number of puzzles are included which relate to mathematical recreations or which led to the development of specific recreations, but the recreations themselves are not treated. Readers interested in mathematical recreations will find references to many of the best-known and most readily available sources in the bibliography.

Finally, manipulative puzzles requiring some kind of apparatus also deserve a book-length treatment of their own, and are excluded here. All the puzzles in this book can be tackled either mentally, or with the assistance of at most pencil and paper and perhaps a few counters.

Compiling this book has taken me back to the days when I was Puzzle Editor of *Games & Puzzles* magazine, and work was a pleasure hard to distinguish from play. I hope that readers will find some of that pleasure in the immense variety of puzzles assembled here.

I shall be happy to receive readers' opinions and suggestions, though I cannot guarantee to respond to every letter personally. I wish you happy and successful puzzling!

D.W.   1992

The Puzzles

The World's Oldest Puzzle

1. There are seven houses each containing seven cats. Each cat kills seven mice and each mouse would have eaten seven ears of spelt. Each ear of spelt would have produced seven hekats of grain. What is the total of all these?

This puzzle, freely paraphrased here, is problem 79 in the Rhind papyrus, our richest source for ancient Egyptian mathematics, which is named after the Scottish Egyptologist A. Henry Rhind, who purchased it in 1858 in Luxor.

The Rhind papyrus is in the form of a scroll about eighteen and a half feet long and thirteen inches wide, written on both sides. It dates from about 1650 BC. The scribe's name was Ahmes, and he states that he is copying a work written two centuries earlier, so the original of the Rhind papyrus was written in the same period as another famous source of Egyptian mathematics, the Moscow papyrus, dating from 1850 BC.

Returning to the cats and mice, about 2800 years after Ahmes, Fibonacci in his *Liber Abaci* (1202) posed this puzzle:

2. Seven old women are travelling to Rome, and each has seven mules. On each mule there are seven sacks, in each sack there are seven loaves of bread, in each loaf there are seven knives, and each knife has seven sheaths. The question is to find the total of all of them.

The resemblance is so strong that surely Fibonacci's problem is a direct descendant, along an historical path that we can no longer trace, of the Rhind puzzle? Not necessarily. There is an undoubted fascination with geometrical series, and the number 7 is not only as magical and mysterious as any number can be, but was especially easy for the Egyptians to handle, because they multiplied by repeated doubling, and $7 = 1 + 2 + 4$. Put these factors together, and you naturally arrive at two similar puzzles.

The St Ives Riddle

3.
> As I was going to St Ives,
> I met a man with seven wives.
> Every wife had seven sacks,
> Every sack had seven cats,
> Every cat had seven kits;
> Kits, cats, sacks and wives,
> How many were going to St Ives?

This rhyme appears in the eighteenth-century *Mother Goose* collection. Is it also descended from the Rhind papyrus and Fibonacci?

Egyptian Fractions

The Egyptians could easily handle simple fractions, but with one remarkable peculiarity. The only fractions they used were $\frac{2}{3}$ and the reciprocals of the integers, the so-called *unit fractions* with unit numerators.

The Rhind papyrus contains a table of fractions in the form $2/n$ for all odd values of n from 5 to 101. They also had a rule for expressing $\frac{2}{3}$ of a unit fraction as the sum of unit fractions: to find $\frac{2}{3}$ of $\frac{1}{5}$, multiply 5 by 2 and by 6: $\frac{2}{3}$ of $\frac{1}{5} = \frac{1}{10} + \frac{1}{30}$. Similarly, $\frac{2}{3}$ of $\frac{1}{8}$ is $\frac{1}{16} + \frac{1}{48}$. Curious though this treatment of fractions may seem to us, no doubt it seemed both natural and easy to them.

Thus their answer to the problem, 'divide seven loaves among ten men' was not 7/10 of a loaf each, but the fraction $\frac{1}{2} + \frac{1}{5}$.

Can all proper fractions be expressed as the sum of unit fractions, without repetition? Yes, as Fibonacci showed, also in his *Liber Abaci*, where he described what is now called the *greedy algorithm*. Subtract the largest possible unit fraction, then do the same again, and so on. Sylvester proved in 1880 that applying this greedy algorithm to the fraction p/q, where p is less than q, produces a sequence of no more than p unit fractions.

4. The greedy algorithm does not work so well if we add the

condition that all the denominators must be odd. There are just five ways to represent 1 as the sum of the smallest possible number of Egyptian fractions, with odd denominators. Which has the smallest largest denominator?

5. What is the smallest fraction $3/n$ for which the greedy algorithm produces a sum in three terms, but two terms are actually sufficient?

The sum of the series $1 + 1/2^2 + 1/3^2 + 1/4^2 \ldots = \pi^2/6$, so the sum of different Egyptian fractions whose denominators are squares cannot exceed $\pi^2/6$, but might equal, for example, $\frac{1}{2}$.

6. How can $\frac{1}{2}$ be represented as the sum of unit fractions with square denominators, with no denominator greater than 35^2?

Think of a Number

7. Problem 29 of the Rhind papyrus is not quite so clear, but it is plausibly the first ever 'Think of a Number' problem. It reads, 'Two-thirds is to be added. One-third is to be subtracted. There remains 10.' In clearer language that reads: 'I think of a number, and add to it two-thirds of the number. I then subtract one-third of the sum. My answer is 10. What number did I think of?'

8. 'If the scribe says to thee, "10 has become $\frac{2}{3} + \frac{1}{10}$ of what?"' is the Egyptian way of saying, in effect, 'I think of a number. Two-thirds of the number plus its tenth make 10. What was the number?'

9. 'A number, plus its two-thirds, and plus its half, plus its seventh, makes 37. What is the number?'

Readers will naturally wish to express the answer in Egyptian fractions!

Sharing the Loaves

Arithmetic progressions have not been as popular in the history of puzzles as geometric ones. There is after all something impressive, mysterious even, in the rapidity with which geometric progressions increase, while arithmetic progressions just plod along, step by equal step.

Yet puzzles about arithmetical progressions can be thought-provoking, as this example illustrates.

10. 'A hundred loaves to five men, one-seventh of the three first men to the two last.'

The meaning is: 'Divide 100 loaves between five men so that the shares are in arithmetical progression, and the sum of the two smaller shares is one-seventh of the sum of the three greatest.'

Squares Without Pythagoras

It is a well-known 'fact' that the ancient Egyptians used knotted ropes to make a 3–4–5 triangle and hence construct accurate right-angles. This 'fact' is actually a myth, based on a suggestion by the historian Moritz Cantor that the Egyptians *might just possibly* have made right-angles this way. There is no evidence that they did anything of the sort, or that they had any knowledge whatsoever of Pythagoras's theorem. They did, however, consider problems about areas and square numbers. This is from the Berlin papyrus:

11. 'If it is said to thee . . . the area of a square of 100 is equal to that of two smaller squares. The side of one is $\frac{1}{2} + \frac{1}{4}$ the side of the other. Let me know the sides of the two unknown squares.'

The Babylonians

Babylonian mathematics was arithmetical and algebraic and far in advance of Egyptian mathematics of the same period. They could solve all the problems in the Rhind papyrus and many more besides.

The Babylonians counted in a sexagesimal system. Instead of counting in tens and hundreds and using tenths and hundredths, and so on, they used multiples of 60, so 6,30 means $6 + (30/60)$, or $6\frac{1}{2}$, and 11,22,30 means $11 + (22/60) + (30/3600)$, or $11\frac{3}{8}$.

Dividing a Field

12. A triangular field is to be divided between six brothers by equidistant lines parallel to one side. The length of the marked side is 6,30 and the area is 11,22,30. What is the difference between the brothers' shares?

This problem is much like Problem 10, which required the construction of an arithmetical series to fit given conditions. Other problems were far more advanced. Thus a tablet from about 1600 BC, contempor-

ary with the Rhind papyrus, leads in modern notation to the solution
of two equations of the form:

$$xy = a \qquad \frac{bx^2}{y} + \frac{cy^2}{x} + d = 0$$

which leads to an equation in x^6, x^3 and a constant.

13. This is from about 1800 BC:

'An area A, consisting of the sum of two squares, is 1000. The side
of one square is 10 less than two-thirds of the other square. What are
the sides of the squares?'

Pythagorean Triples

The Babylonians, unlike the Egyptians, not only knew Pythagoras's
theorem, but they were also familiar with Pythagorean triples, triples
of whole numbers such as 3–4–5 which are the sides of right-angled
triangles. Their investigations of Pythagorean triples started a trail of

discovery, leading through Diophantus to Fermat, to the present day.

14. Ladders were a natural source of problems. A ladder of length
0,30 is standing upright against a wall. If the upper end slides down
the wall a distance of 0,6, how far will the lower end move out from
the wall?

'Plimpton 322' is the name of a clay tablet dating from between 1900 BC
and 1600 BC. It contains fifteen numbered lines with two figures in each
line which are the hypotenuse and one leg of a right-angled triangle.

Although the lengths given seem to vary in an apparently irregular
way from one line to the next, in fact their *ratios* increase steadily
from $169/119 = 1.42$ in the first line to $106/56 = 1.89$ in the last.

15. Problem: find the hypotenuse and one leg of a right-angled
triangle whose ratio is approximately 1.54.

The Greeks

Archimedes' Cattle Problem

Archimedes (287–212 BC) was the greatest mathematician of antiquity,
a wonderful geometer who anticipated the calculus, invented hydro-

statics, and studied giant numbers in his book *The Sandreckoner*. It is a curiosity that one extremely difficult problem and one simple recreation are associated with his name.

16. 'If thou art diligent and wise, O stranger, compute the number of cattle of the Sun, who once upon a time grazed on the fields of the Thrinician isle of Sicily, divided into four herds of different colours, one milk white, another glossy black, the third yellow and the last dappled. In each herd were bulls, mighty in number according to these proportions: understand, stranger, that the white bulls were equal to a half and a third of the black together with the whole of the yellow, while the black were equal to the fourth part of the dappled and a fifth, together with, once more, the whole of the yellow. Observe further that the remaining bulls, the dappled, were equal to a sixth part of the white and a seventh, together with all the yellow. These were the proportions of the cows: the white were precisely equal to the third part and a fourth of the whole herd of the black; while the black were equal to the fourth part once more of the dappled and with it a fifth part, when all, including the bulls, went to pasture together. Now the dappled in four parts were equal in number to a fifth part and a sixth of the yellow herd. Finally the yellow were in number equal to a sixth part and seventh of the white herd. If thou canst accurately tell, O stranger, the number of cattle of the Sun, giving separately the number of well-fed bulls and again the number of females according to each colour, thou wouldst not be called unskilled or ignorant of numbers, but not yet shalt thou be numbered among the wise . . .

'But come, understand also all these conditions regarding the cows of the Sun. When the white bulls mingled their number with the black, they stood firm, equal in depth and breadth, and the plains of Thrinacia, stretching far in all ways, were filled with their multitude. Again, when the yellow and the dappled bulls were gathered into one herd they stood in such a manner that their number, beginning from one, grew slowly greater till it completed a triangular figure, there being no bulls of other colours in their midst nor none of them lacking.

'If thou art able, O stranger, to find out all these things and gather them together in your mind, giving all the relations, thou shalt depart crowned with glory and knowing that thou hast been adjudged perfect in this species of wisdom.'

Archimedes' cattle problem is extant in more than one manuscript.

The 'most complete' version contains the extra conditions that follow the ellipsis. These conditions are ambiguous: because the bulls are longer than they are broad, the condition that the white and black bulls together form a square does not necessarily mean that their total is a square number; it could be merely a rectangular number.

It is plausible that the more difficult interpretation is intended. Archimedes dedicated the problem to his friend the great Alexandrian astronomer Eratosthenes, which suggests that it was extremely difficult, and Archimedes' interest in very large numbers is evident from his *Sandreckoner*, in which he calculated the number of grains of sand needed to fill a sphere whose centre was the centre of the earth and which extended to reach the sun. Also, in classical antiquity a difficult problem was often described as a *problema bovinum* or a *problema Archimedis*, such was his fame. If this is so, then the solution is indeed complex and extraordinarily lengthy. A. Amthor calculated in 1880 that the total number of cattle in this case is a number of 206,545 digits. Further details will be found in Sir Thomas Heath's *A History of Greek Mathematics*, p. 319.

If, however, the latter conditions are ignored, and the reader is willing to be judged merely 'not unskilled' in the art, rather than perfectly wise, then the answer will be found in the Solutions section.

Loculus of Archimedes

Several ancient sources refer to this puzzle, which is described in an Arabic manuscript, *The Book of Archimedes on the Division of the Figure Stomaschion*.

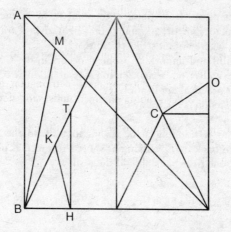

The Loculus consists of fourteen pieces making a square. The method of division is almost self-evident: M, T and C are mid-points, and HK passes through A and OC through B.

The object of the puzzle is to make figures with these pieces. Unlike the Chinese Tangram puzzle, which might be said to have too few pieces, this has rather a lot. (Is it a coincidence that the Tangram has the magical number of seven pieces and the Loculus exactly twice as many?)

17. How can this figure of an elephant be composed from the pieces of the Loculus?

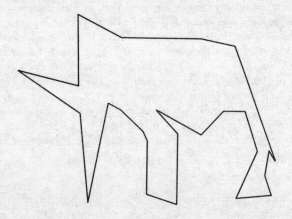

Light Reflected off a Mirror

A ray of light passes from point A to point B, by bouncing off the surface of a plane mirror. Assuming that light always travels by the shortest path, where does it strike the mirror?

This beautiful and important problem occurs in the *Catoptrica* of Heron of Alexandria (*c*.75 AD). Heron's assumption is correct, so it has important practical applications.

A modern version is the following:

18. Mary, who is standing at S, wishes to walk to the river for a drink and then back to T, walking as short a distance as possible. To what point on the river bank should she walk?

•T

•S

19. From the *Greek Anthology*, *c.* 500 AD: 'I am a brazen lion; my spouts are my two eyes, my mouth and the flat of my right foot. My right eye fills a jar in two days, my left eye in three, and my foot in four. My mouth is capable of filling it in six hours; tell me how long all four together will take to fill it?'

Heron was a master of mechanical devices. His *Pneumatica* describes scores of machines operated by wind and water, so it is no surprise that the famous cistern problem occurs in his *Metrika*.

Famous? This is the infamous problem about the tank which is filled with water from several pipes, which was still being used to torture schoolchildren till the middle of this century, and which has become a byword for 'useless' mathematics. This is a great pity, because the idea behind it is far from useless and turns up in many important situations.

20. From *The Tutorial Arithmetic* by W. P. Workman, published in 1920: 'A and B together can do a piece of work in 6 days, B and C together in 20 days, C and A together in $7\frac{1}{2}$ days. How long will each require separately to do the same work?'

Heron was also a geometrician:

21. Find two rectangles, with integral sides, such that the area of the first is three times the area of the second, and the perimeter of the second is three times the perimeter of the first.

22. In a right-angled triangle with integral sides, the sum of the area and the perimeter is 280. Find the sides and the area.

The First Pure Number Puzzles

23. What number must be added to 100 and to 20 (the same number to each) so that the sums are in the ratio 3:1?

24. Two numbers are such that if the first receives 30 from the second, they are in the ratio 2:1, but if the second receives 50 from the first, their ratio is then 1:3. What are the numbers?

25. The sums of four numbers, omitting each of the numbers in turn, are 22, 24, 27 and 20, respectively. What are the numbers?

These problems are nos. 8, 15 and 17 of Book I of the *Arithmetica* of Diophantos of Alexandria (*c.* 250 AD). Typically, the solutions are all whole numbers. All his problems concern integers or rational numbers, and such problems in integers are named *Diophantine* after him. While studying such problems he is led to discuss the multiplication of positive and negative numbers. Coincidentally, a commentary on his work was written by Hypatia (*c.* 410), the first known woman mathematician, who was murdered by a Christian mob in the year 415.

The works of Diophantos vary from the simply puzzling and puzzlingly simple, to very difficult questions which had a stunning impact when his works were first translated into Latin and studied by European mathematicians more than 1200 years later. Xylander wrote in 1575: 'I came to believe that in Arithmetic and Logistic "I was somebody". And in fact by not a few, among them some true scholars, I was adjudged an Arithmetician beyond the common order. But when I first came upon the work of Diophantos, his method and his reasoning so overwhelmed me that I scarcely knew whether to think of my former self with pity or with laughter.'

The elementary problems that Diophantos solves could all have been presented, had he so wished, as puzzles in everyday settings, and were by other writers. Here the numbers themselves are personified:

26. 'To find three numbers such that, if each give to the next following a given fraction of itself, in order, the results after each has given and taken may be equal.

'Let the first give $\frac{1}{3}$ of itself to the second, the second give $\frac{1}{4}$ of itself to the third, and the third give $\frac{1}{5}$ of itself to the first. What are the numbers?' (Diophantos assumes that all these transactions take place simultaneously, and not in sequence.)

Square Problems

27. Find three numbers such that the product of any two added to the third gives a square.

28. Find three numbers such that their sum is a square and the sum of any pair is a square.

29. 'A man buys a certain number of measures of wine, some at 8 drachmas, some at 5 drachmas each. He pays for them a square number of drachmas; and if we add 60 to this number, the result is a square, the side of which is equal to the whole number of measures. Find how many he bought at each price.'

The Area Enclosed Against the Seashore

> So they reached the place where you will now behold mighty walls and the rising towers of the new town of Carthage; and they bought a plot of ground named Byrsa . . . for they were to have as much as they could enclose with a bull's hide.
>
> Virgil, *Aeneid*, Book I, ll. 360–70

Questions and facts about extremes have a natural attraction; witness the runaway success of the *Guinness Book of Records*. At a more serious level, many scientific principles can be expressed in terms of maxima and minima, as Heron's problem of the ray of light reflecting off a mirror illustrates (see p. 10).

30. 'Given a long string, with which to enclose the maximum possible area against a straight shore-line, how should the string be disposed?'

Here are two variants:

31. This frame is composed of four rods that are hinged to each other at their ends. When will the area enclosed by the frame be a maximum?

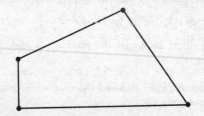

32. This figure shows the corner of a room with a screen, composed of two identical halves hinged together, placed to cut off a portion of the corner of the room. How should the screen be placed to enclose as large an area as possible?

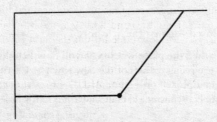

33. An isosceles triangle has two equal sides of length 10, hinged together. What is the maximum area of the triangle?

Metrodorus and the *Greek Anthology*

The *Greek Anthology* is a collection of literary verses and epigrams. Surprisingly, Book XIV comprises a large number of riddles, enigmas and puzzles, credited to Metrodorus (*c.* 500 AD).

These include not only arithmetical puzzles, but very early word puzzles, including beheadings, in which a word loses letter after letter from its front end but always remains a proper word, and this puzzle:

> If you put one hundred in the middle of a burning fire, you will find the son and a slayer of a virgin.

The answer is to put the Greek symbol for 100, *rho*, into the word for fire, *pyros*, to get Pyrrhos, the son of Deidamia and the slayer of Polyxena.

The arithmetical and logical puzzles include what were already classic problems, such as finding the weights of bowls given in arithmetical progression, and the cisterns problem, and new types:

> My father-in-law killed my husband and my husband killed my father-in-law; my brother-in-law killed my father-in-law, and my father-in-law my father.

The answer is Andromache. Achilles, father of her second husband, Pyrrhus, killed Hector, Pyrrhus killed Priam, Paris killed Achilles, and Achilles killed her father, Eetion.

34. ' "Best of clocks, how much of the day is past?" There remains twice two-thirds of what is gone.' (Problem 6; the day is counted as lasting for 12 hours.)

35. 'This tomb holds Diophantos. Ah, how great a marvel! the tomb tells scientifically the measure of his life. God granted him to be a boy for the sixth part of his life, and adding a twelfth part to this, He clothed his cheeks with down; He lit him the light of wedlock after a seventh part, and five years after his marriage He granted him a son. Alas! late-born wretched child; after attaining the measure of half his father's life, chill Fate took him. After consoling his grief by this science of numbers for four years he ended his life.' (Problem 126)

36. 'I desire my two sons to receive the thousand staters of which I am possessed, but let the fifth part of the legitimate one's share exceed by ten the fourth part of what falls to the illegitimate one.' (Problem 11)

Arabic Puzzles

Al-Khwarizmi (*c.* 825 AD)

Al-Khwarizmi wrote a book, *al-Kitab al-mukhtasar hisab al-jabr wa'l-muqabala*, or *The Compendious Book on Calculations by Completion and Balancing*, on the solution of equations. Later Arabic works tended to use the same expression *al-jabr wa'l-muqabala*, or just *al-jabr*, to refer to books on the same theme, from whence we eventually derive our word 'algebra'.

The second half of the same book deals with problems of inheritance, according to Islamic law. This is an essential study for Islamic jurists, as it had previously been for Roman lawyers, though as Ibn Khaldun wrote in the fourteenth century, 'Some authors are inclined to exaggerate the mathematical side of the discipline and to pose problems requiring for their solution various branches of arithmetic, such as algebra, the use of roots, and similar things' (Berggren, 1986, p. 53).

Well, mathematicians would, wouldn't they! This problem is practical:

37. A woman dies, leaving her husband, a son and three daughters. She also leaves $\frac{1}{8} + \frac{1}{7}$ of her estate to a stranger. According to law, the husband receives one quarter of the estate and the son receives double the share of a daughter, but this division is made only after the legacy to the stranger has been paid. How must the inheritance be divided?

Abul Wafa (940–998)

Abul Wafa was born in Buzjan in Persia in 940. He wrote commentaries on Euclid and Diophantos and Al-Khwarizmi, but he is best known for his study of geometrical dissections and of constructions with a rusty compass, meaning a compass which is so stiff that it can be used with only one opening.

38. Construct an equilateral triangle inside a square, so that one vertex is at a corner of the square and the other two vertices are on the opposite sides.

39. **Three Squares into One**   Dissect three equal squares into one square.

40. Dissect two identical larger squares plus one smaller square into one square.

41. How can two regular hexagons, of different sizes, be dissected into seven pieces which fit together to make one, larger, regular hexagon?

42. Given three identical triangles, and one smaller triangle similar to them in shape, how can all four be dissected into one triangle?

43. **The Rusty Compass**   Using only a straight-edge and a compass with a fixed opening, construct at the endpoint A of a segment AB a

perpendicular to that segment, without prolonging the segment beyond A.

44. Using only a straight-edge and fixed-opening compasses, divide a given line-segment into any given number of equal parts.

45. Construct a regular pentagon in a given circle, using only a straight-edge and a compass with a fixed opening equal to the radius of the circle.

Sissa and the Chessboard

Ibn Kallikan (*c.* 1256) was the first author to tell the story of Sissa ben Dahir, who was asked by the Indian King Shirham what he desired as a reward for inventing the game of chess:

46. '"Majesty, give me a grain of wheat to place on the first square, and two grains of wheat to place on the second square, and four grains of wheat to place on the third, and eight grains of wheat to place on the fourth, and so, Oh King, let me cover each of the sixty-four squares on the board."

'"And is that all you wish, Sissa, you fool?" exclaimed the astonished King.

'"Oh, Sire," Sissa replied, "I have asked for more wheat than you have in your entire kingdom, nay, for more wheat than there is in the whole world, verily, for enough to cover the whole surface of the earth to the depth of the twentieth part of a cubit."'

How many grains of wheat did Sissa require?

Indian Puzzles

The Bhakshali manuscript was found in 1881 in north-west India and dates from somewhere between the third and twelfth centuries, depending on which authority you choose. It contains the earliest – if it really dates as early as the third century – version of what came to be called 'One Hundred Fowls' problem (see Problem 74), in this form:

47. Twenty men, women and children earn twenty coins between them. Each man earns 3 coins, each woman $1\frac{1}{2}$ coins and each child $\frac{1}{2}$ coin. How many men, women and children are there?

Mahavira (*c.* 850) wrote on elementary mathematics. Problems 48 to 54 are from his book the *Ganita-Sara-Sangraha*.

48. 'Three *puranas* formed the pay of one man who is a mounted soldier; and at that rate there were sixty-five men in all. Some (among them) broke down, and the amount of their pay was given to those that remained in the field. Of this, each man obtained 10 *puranas*. You tell me, after thinking well, how many remained in the field and how many broke down.'

49. 'Two market-women were selling apples, one at two for 1 cent, and the other at three for 2 cents. They had thirty apples apiece. In order to end their competition they formed a trust, pooling their stock and selling the apples at five for 3 cents. This was to their advantage, since under the new arrangement they took, in total, 36 cents, while under the old system they would have received a total of only 35 cents.

'Two other women, who also had thirty apples apiece, and who were selling them at two for 1 cent and three for 1 cent, also formed a trust to sell their apples, at five for 2 cents. But instead of the total of 25 cents which they would have taken in operating separate enterprises, their trust grossed only 24 cents. Why?'

50. 'One night, in a month of the spring season, a certain young lady ... was lovingly happy along with her husband on ... the floor of a big mansion, white like the moon, and situated in a pleasure-garden with trees bent down with the load of bunches of flowers and fruits, and resonant with the sweet sounds of parrots, cuckoos and bees which were all intoxicated with the honey obtained from the flowers therein. Then on a love-quarrel arising between the husband and the wife, that lady's necklace made up of pearls became sundered and fell on the floor. One-third of that necklace of pearls reached the maid-servant there; one-sixth fell on the bed; then one-half of what remained (and one-half of what remained thereafter and again one-half of what remained thereafter) and so on, counting six times [in all] fell all of them everywhere; and there were found to remain [unscattered] 1,161 pearls; and if you know ... give out the measure of the pearls.'

51. 'In how many ways can different numbers of flavours be used in combination together, being selected from the astringent, the bitter, the sour, the pungent, and the saline, together with the sweet taste?'

52. 'Three merchants saw in the road a purse [containing money]. One said, "If I secure this purse, I shall become twice as rich as both of you together."

'Then the second said, "I shall become three times as rich."

'Then the third said, "I shall become five times as rich."

'What is the value of the money in the purse, as also the money on hand [with each of the three merchants]?'

53. Arrows, if they are thin cylinders, circular in cross-section, can be packed in hexagonal bundles:

'The circumferential arrows are eighteen in number. How many [in all] are the arrows to be found [in the bundle] within the quiver?'

54. Two pillars are of known height. Two strings are tied, one to the top of each. Each of these two strings is stretched so as to touch the foot of the other pillar. From the point where the two strings meet, another string is suspended vertically till it touches the ground. What is the length of this suspended string?

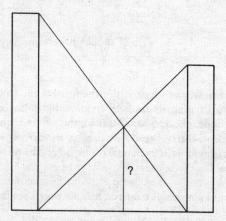

This is identical to puzzles about ladders resting across passageways in which the heights of the points at which they touch are given. If not the vertical heights but the lengths of the ladders are known, then the problem of finding the height of their intersection is far harder. (See p. 131.)

Bhaskara (1115–*c*. 1185) was an astronomer and mathematician whose most famous work, the *Lilavati*, from which the following problems are taken, was addressed to his daughter, or perhaps his wife. It ends with this delightful paragraph, typical of the Indian style of the period:

> Joy and happiness is indeed ever increasing in this world for those who have *Lilavati* clasped to their throats, decorated as the members are with neat reduction of fractions, multiplication and involution, pure and perfect as are the solutions, and tasteful as is the speech which is exemplified.

55. 'In an expedition to seize his enemy's elephants, a king marched 2 *yojanas* the first day. Say, intelligent calculator, with what increasing rate of daily march did he proceed, since he reached his foe's city, a distance of 80 *yojanas*, in a week?'

56. 'A snake's hole is at the foot of a pillar which is 15 cubits high and a peacock is perched on its summit. Seeing a snake, at a distance of thrice the pillar's height, gliding towards his hole, he pounces obliquely upon him. Say quickly at how many cubits from the snake's hole do they meet, both proceeding an equal distance?'

It is natural that Hindu writers should have considered sooner or later the permutations and combinations of the attributes of their gods:

57. 'How many are the variations in the form of the God Siva by the exchange of his ten attributes held reciprocally in his several hands: namely, the rope, the elephant's hook, the serpent, the tabor, the skull, the trident, the bedstead, the dagger, the arrow, and the bow: as those of Vishnu by the exchange of the mace, the discus, and lotus and the conch?'

The final Hindu problem is unattributed, but on a popular theme:

58. The first man has sixteen azure-blue gems, the second has ten emeralds, and the third has eight diamonds. Each among them gives to each of the others two gems of the kind owned by himself; and then all three men come to be possessed of equal wealth. What are the prices of those azure-blue gems, emeralds and diamonds?

Puzzles from China

The First Magic Square

59. How can the numbers 1 to 9 be arranged in the cells of this square so that the sums of every row and column and both diagonals are equal?

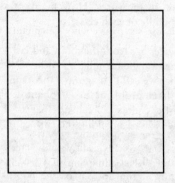

The resulting figure has essentially the arrangement of the *Lo Shu*, which, in Chinese legend going back at least to the fifth century BC, was the gift of a turtle from the River Lo to the Emperor Yu the Great, who first controlled the flow of the Lo and the Yellow rivers.

The Nine Chapters

The Nine Chapters of Mathematical Art is supposed to have been written in the third century BC, and contains the first known examples of the solution of linear simultaneous equations, well ahead of the West, as well as the extraction of square and cube roots.

60. 'Suppose that there are a number of rabbits and pheasants confined in a cage, in all thirty-five heads and ninety-four feet; required the number of each?'

61. 'A number of men bought a number of articles, neither of which are known; it is only known that if each man paid 8 cash, there would be a surplus of 3 cash; and if each man paid 7 cash, there would be a deficiency of 4 cash. Required the respective numbers?'

62. 'If five oxen and two sheep cost 10 taels of gold, and two oxen and five sheep cost 8 taels, what are the prices of the oxen and sheep respectively?'

63. 'There are three classes of corn, of which three bundles of the first class, two of the second class and one of the third make 39 measures. Two of the first, three of the second and one of the third make 34 measures. And one of the first, two of the second and three of the third make 26 measures. How many measures of grain are contained in one bundle of each class?'

The following puzzles are from the ninth and last section of the book, and all concern right-angled triangles and the *Gougu* theorem, as the Chinese called what we call Pythagoras's theorem.

In contrast to later problems in Diophantos, these are all set in remarkably realistic contexts, realistic that is if a mathematician happened to notice a reed breaking the surface of a pool, or a chain hanging from a pillar.

64. 'There is a pool 10 feet square, with a reed growing vertically in the centre, its roots at the bottom of the pool, which rises a foot above the surface; when drawn towards the shore it reaches exactly to the brink of the pool; what is the depth of the water?'

65. 'A chain suspended from an upright post has a length of 2 feet lying on the ground, and on being drawn out to its full length, so as just to touch the ground, the end is found to be 8 feet from the post; what is the length of the chain?'

The following problem was also presented by the Indian mathematician and astronomer Brahmagupta, more than 600 years later:

66. 'There is a bamboo 10 feet high, the upper end of which being broken down on reaching the ground, the tip is just 3 feet from the stem; what is the height of the break?'

67. 'What is the largest circle that can be inscribed within a right-angled triangle, the two short sides of which are respectively 8 and 15?'

68. 'Of two water weeds, one grows 3 feet and the other 1 foot on the first day. The growth of the first becomes every day half of that of the preceding day, while the other grows twice as much as on the day before. In how many days will the two grow to equal heights?'

Sun Tsu Suan-Ching (fourth century AD)

69. 'A woman was washing dishes in a river, when an official whose business was overseeing the waters demanded of her: "Why are there so many dishes here?"

'"Because a feasting was entertained in the house," the woman replied. Thereupon the official inquired the number of guests.

'"I don't know," the woman said, "how many guests there had been; but every two used a dish for rice between them; every three a dish for broth; every four a dish for meat; and there were sixty-five dishes in all."'

The next problem is an example of the famous Chinese Remainder Theorem. Such problems had practical applications to calendar problems, when cycles of different lengths are compared.

70. 'There are certain things whose number is unknown. Repeatedly divided by 3, the remainder is 2; by 5 the remainder is 3; and by 7 the remainder is 2. What will be the number?'

71. 'There are three sisters, of whom the eldest comes home once every five days, the middle in every four days, and the youngest in every three days. In how many days will all the three meet together?'

Liu Hui (263 AD), in the *Hai Tao Suan-Ching*, or *Sea-Island Arithmetical Classic*, poses this simple puzzle:

72. What is the size of a square inscribed in the corner of a right-angled triangle to touch the hypotenuse?

The *Chang Sh'iu-Chien Suan-Ching*, or *The Arithmetical Classic of Ch-iu Chien* (sixth century), poses one of the earliest chasing and returning puzzles:

73. 'A man, who had stolen a horse, rode away on its back. When he had gone 37 miles, the owner discovered the theft and pursued the thief for 145 miles; he then returned, [believing himself] unable to overtake him. When he turned back the thief was riding 23 miles ahead of him; if he had continued in his pursuit without coming back, in how many further miles would he have overtaken him?'

It also contains the earliest '100 fowls' problem:

74. 100 fowls are sold for 100 shillings, the cocks being sold for 5 shillings each, the hens for 3 shillings and the chicks for $\frac{1}{3}$ shilling each. How many of each were sold?

Yang Hui (*c.* 1270 AD) wrote an 'Arithmetic in Nine Sections', which contains the very first extant representation of what we in the West call Pascal's Triangle (from an earlier Chinese source, *c.* 1000 AD). His book was called, apparently, *Hsu Ku Chai Chi Suan Fa* (1275). It contains the following magic configuration:

75. Arrange the numbers 1 to 33 in these circles so that every circle and every diameter has the same total.

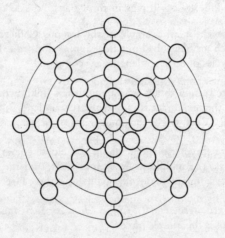

'Propositions to Sharpen Up the Young'

Propositiones ad acuendos juvenes was written in the monastery of Augsberg, about the year 1000, and has been included in the works of Alcuin (*c.* 732–804), the English scholar and churchman who spent his life at the court of the Emperor Charlemagne, on the grounds that Alcuin writes in one of his letters to the Emperor that he is sending him, among other matters, 'certain subtle figures of arithmetic, for pleasure', and this might be that collection.

Anyway, it is the earliest European collection of mathematical and logical puzzles, and contains the first appearance of many well-known puzzle types.

Like the problems of Metrodorus in the *Greek Anthology*, a few riddles and trick questions find their way into the fifty-three problems in the collection.

76. 'An ox ploughs a field all day. How many footprints does he leave in the last furrow?' (Problem XIV)

'A man has 300 pigs, and orders that the pigs must be killed, an odd number each day, in three days. Say how many pigs must be killed each day.'

This is Problem XLIII. The answer is: 'This is a fable. Nobody can solve how to kill 300 or 30 pigs in three days, an odd number each day. This puzzle is given to children to solve.'

This could be cruelty to little children, but it is also an early recognition that some problems simply cannot be solved.

77. 'Two wholesalers with 100 shillings between them bought some pigs with the money. They bought at the rate of five pigs for 2 shillings, intending to fatten them up and sell them again, making a profit. But when they found that it was not the right time of year for fattening pigs, and they were not able to feed them through the winter, they tried to sell them again to make a profit. But they couldn't, because they could only sell them for the price they had paid for them ... When they saw this, they said to each other: "let's divide them". By dividing them, and selling them at the rate they had bought them for, they made a profit. How many pigs were there, and how could they be divided to make a profit, which could not be made by selling them all at once?' (Problem VI)

78. 'A king ordered his servant to collect an army from thirty manors, in such a way that from each manor he would take the same number of men as he had collected up to then. The servant went to the first manor alone; to the second he went with one other ...' How many men were collected in all? (Problem XIII)

79. 'If two men each take the other's sister in marriage, what is the relationship between their sons?' (Problem XI)

80. 'A father, when dying, gave to his sons thirty glass flasks, of which ten were full of wine, ten were half full, and the last ten were

empty. Divide the [wine] and the flasks, so that each of the three sons receives equally of both glass and wine.' (Problem XII)

The next three classic problems, like the last two, appear for the first time in Alcuin.

81. Three Friends and their Sisters 'Three men, each with a sister, needed to cross a river. Each one of them coveted the sister of another. At the river, they found only a small boat, in which only two of them could cross at a time. How did they cross the river, without any of the women being defiled by the men?' (Problem XVII)

82. A Man, a Goat, and a Wolf 'A man takes a wolf, a goat and a cabbage across the river. The only boat he could find could take only two of them at a time. But he had been ordered to transfer all of these to the other side in good condition. How could this be done?' (Problem XVIII)

83. A Very Heavy Man and Woman 'A man and a woman, each the weight of a loaded cart, with two children who between them weigh as much as a loaded cart, have to cross a river. They find a boat which can only take one cartload. Make the transfer, if you can, without sinking the boat.' (Problem XIX)

84. 'A dying man left 960 shillings and a pregnant wife. He directed that if a boy was born, he should receive three-quarters of the whole, and the child's mother should receive one quarter. But if a daughter was born, she would receive seven-twelfths, and her mother five-twelfths. It happened however that twins were born – a boy and a girl. How much should the mother receive, how much the son, and how much the daughter?' (Problem XXV)

85. 'A stairway consists of 100 steps. On the first step stands a pigeon; on the second, two pigeons; on the third, three; on the fourth, four; on the fifth, five; and so on every step up to the hundredth. How many pigeons are there altogether?' (Problem XLII)

Liber Abaci

Leonardo of Pisa (*c.* 1175–1250) was a member of the Bonacci family, so was often known as Fibonacci (*filio Bonacci*). As a young man he travelled to Bugia in North Africa to help his father, who directed a trading post there, and learnt from the local Arabs the new Indian numerals, our Hindu numerals, which he helped to introduce to Europe.

He wrote a book on calculations, the *Liber Abaci*, a compendium on geometry and trigonometry, *Practica geometriae*, and the *Liber quadratorum* (*The Book of Squares*) on Diophantine problems.

He describes in the prologue to *The Book of Squares* how he was invited to the court of Emperor Frederick II of Sicily to compete in a mathematical tournament. He solved all three problems posed to him by John of Palermo. The first, in the style of Diophantos, was to:

86. 'Find a rational number such that 5 added to, or subtracted from, its square, is also a square.'

The second was to solve the cubic

$$x^3 + 2x^2 + 10x = 20$$

Leonardo found the solution, 1.3688081075, which is correct to nine decimal places.

This is the third problem:

87. 'Three men possess a pile of money, their shares being 1/2, 1/3, 1/6. Each man takes some money from the pile until nothing is left. The first man returns 1/2 of what he took, the second 1/3 and the third 1/6. When the total so returned is divided equally among the men it is found that each then possesses what he is entitled to. How much money was in the original pile, and how much did each man take from the pile?'

Breeding Rabbits

Fibonacci is best remembered for the following problem, which leads to the Fibonacci sequence:

88. 'A certain man put a pair of rabbits in a place surrounded on all sides by a wall. How many pairs of rabbits can be produced from

that pair in a year if it is supposed that every month each pair begets a new pair which from the second month on becomes productive?'

89. 'A lion would take four hours to eat one sheep; a leopard would take five hours; and a bear would take six; we are asked, if a single sheep were to be thrown to them, how many hours would they take to devour it?'

90. 'A man left to his oldest son one bezant and a seventh of what was left; then, from the remainder, to his next son he left two bezants and a seventh of what was left; then, from the new remainder, to his third son he left three bezants and a seventh of what was left. He continued in this way, giving each son one bezant more than the previous son and a seventh of what remained. By this division it developed that the last son received all that was left and all the sons shared equally. How many sons were there and how large was the man's estate?'

91. 'A man entered an orchard [with] seven gates, and there took a certain number of apples. When he left the orchard he gave the first guard half the apples and one apple more. To the second guard he gave one half of his remaining apples and one apple more. He did the same to each of the remaining five guards, and left the orchard with one apple. How many apples did he gather in the orchard?'

•

92. Serpent Climbing out of a Well   Dell'Abaco (*c.* 1370) discusses this famous puzzle. A serpent lies at the bottom of a well whose depth is 30. It starts to climb, rising up $\frac{2}{3}$ every day and falling back $\frac{1}{3}$ at night. How long does it take to climb out of the well?

93. The Best View of a Statue   From what distance will a statue on a plinth subtend the largest angle? (See figure opposite.)

If you are too close, the statue will appear greatly foreshortened, but if you walk back too far, it will just appear small.

This problem was originally posed by Regiomontanus (1436–76) in 1471 to Christian Roder, as a question about a suspended vertical rod. It is notable as the first extremal problem since the days of antiquity and Heron's problem about the ray of light bouncing off a mirror.

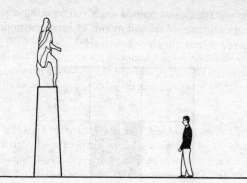

The same problem has been re-invented many times, most recently in this practical form:

According to the rules of rugby union football, a conversion of a try must be taken on a line extending backwards from the point of touchdown, at right-angles to the goal-line. From which point on this line should the conversion be taken, if the aim is to maximize the angle subtended by the goal-posts? This problem applies only when the try is not scored between the posts.

The Couriers Meeting

The *Treviso Arithmetic* (1478) poses this problem:

94. 'The Holy Father sent a courier from Rome to Venice, commanding him that he reach Venice in seven days. And the most illustrious Signoria of Venice also sent another courier to Rome, who should reach Rome in nine days. And from Rome to Venice is 250 miles. It happened that by order of these lords the couriers started on their journeys at the same time. It is required to find in how many days they will meet.'

The Nuns in their Cells

Pacioli in his *De Viribus* (c. 1500) posed this problem of rearrangements:

95. There are eight nuns, one in each cell, making a total of three nuns along each side of the courtyard. How can they be rearranged so that there are four nuns along each side?

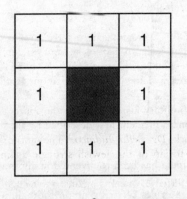

•

Nicolas Chuquet was a doctor by profession, and also the best French mathematician of his time. These two problems are from his *Triparty en la science des nombres*, published in 1484.

96. A carpenter agrees to work on the condition that he is paid £2 for every day that he works, while he forfeits £3 every day that he does not work. At the end of thirty days he finds he has paid out exactly as much as he received. How many days did he work?

97. **Liquid Pouring** This problem first appears in Chuquet. You have two jars holding 5 and 3 pints respectively, neither jar being marked in any way. How can you measure exactly 4 pints from a cask, given that you are allowed to pour liquid back into the cask?

The Josephus Problem

Chuquet was also the first to present an incident in the life of Josephus as a problem:

98. Josephus, during the sack of the city of Jotapata by the Emperor Vespasian, hid in a cellar with forty other Jews who were determined to commit suicide rather than fall into the hands of the Romans. Not wishing to abandon life, he proposed that they form a circle and that every third person, counting round the circle, should die, in the order

in which they were selected. In other words, the count was: 'One, two, three *out*, four, five, six *out* . . .' Where did he place himself, and a companion who also wished to live, in order to ensure that they were the last two remaining?

When Roman troops were judged to have shown cowardice, they were lined up and every tenth man picked out for summary execution, whence our expression 'to decimate' (which, however, has come to possess the stronger meaning of 'reducing to one tenth' of the original number).

This snatch of military history may be the source for the Josephus problem, based on an incident first described by the unknown author of the early work *De Bello Iudaico*. The author describes how Josephus, the historian of the Jewish struggle against the Romans, once saved himself by just this trick.

Later versions pitted Christians against their enemy of the period, the Turks:

99. On board a ship, tossed by storms and in danger of shipwreck, are fifteen Christians and fifteen Turks. To lighten the load and save the ship, half are to be thrown overboard. One of the Christians suggests that all should stand in a circle and every ninth person counting round the circle should be chosen. How should the Christians arrange themselves in the circle to ensure that only the Turks die?

100. In a later Japanese version, thirty children, sons of the same father by his first and second marriages, are too numerous to share his inheritance. So the second wife suggests that the children be placed in a circle and eliminated by counting continually around the circle, eliminating every *n*th child.

By malice, she ensures that the first fourteen children to be eliminated are all the sons of the first wife. The remaining child of that wife, seeing that he is alone, suggests that the order of counting now be reversed, and the second wife agrees, confident that one of her children must be the last survivor, but to her mortification, all her own children are then eliminated.

How were the children arranged, and how was the count done?

Exchanging the Knights

This is one of the earliest recreative chess problems, posed by Guarini di Forli in 1512.

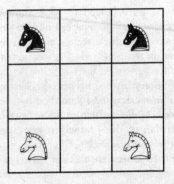

101. Two white knights and two black knights are placed at the opposite corners of this portion of a chessboard. How can the white knights take the places of the black knights, and vice versa, moving according to the rules of chess?

•

Niccolò Fontana (c. 1499–1557), nicknamed Tartaglia (the Stammerer), was the brilliant mathematician who discovered how to solve the cubic equation, only to have Cardano wheedle the solution out of him and publish it himself.

These problems are from his *General Trattato* of 1556 and *Quesiti et Inventioni Diverse* of 1546.

102. A man has three pheasants that he wishes to give to two fathers and two sons, giving each one pheasant. How can it be done?

103. A man dies, leaving seventeen horses to be divided among his heirs, in the proportions $\frac{1}{2} : \frac{1}{3} : \frac{1}{9}$. How can this be done?

Tartaglia also gives a problem of this type:

104. A dishonest servant removes 3 pints of wine from a barrel, replacing them with water. He repeats his theft twice, removing in total 9 pints and replacing them with water. As a result the wine remaining in the barrel is of half its former strength. How much wine did the barrel originally hold?

Bachet

Claude Gaspar Bachet de Meziriac (1581–1638) was a poet and translator, and one of the earliest members of the Académie Française as well as a mathematician.

He is famous for two works, his Greek text of the *Arithmetica* of Diophantos (1621), accompanied by his own Latin commentary, and the first European work devoted to mathematical recreations, *Problèmes plaisans et délectables qui se font par les nombres* (1612).

The *Problèmes plaisans* was largely a compendium of previous puzzles. It contains river-crossing problems originating with Alcuin, a method of constructing magic squares which is that found in Moschopoulos, the Josephus problem as solved by Tartaglia, a liquid-pouring problem, and several think-of-a-number tricks, which are presented here in the form of problems: how is the original number recovered after the following operations?

105. A person chooses secretly a number, and trebles it, telling you whether the product is odd or even. If it is even, he takes half of it, or if it is odd, he adds one and then takes one half. Next he multiplies the result by 3, and tells you how many times 9 will divide into the answer, ignoring any remainder. The number he chose is – what?

106. The subject chooses a number less than 60 and tells you the remainders when it is divided by 3, 4 and 5, separately, not successively. The original number is – what?

107. The first person secretly chooses a number of counters, greater than 5, and the second person takes three times as many. The first then gives 5 counters to the second, who then gives the first three times as many counters as the first has in his hand. You can now say that the second person has – how many – counters in his hand?

Problèmes plaisans also contains the famous problem of the weights. Weights for use with a balance were traditionally made in nested form, so that one weight fitted inside the other and the largest weight contained all the smaller weights. Modern sets of weights in which each fits snugly into the top of the next in the series are a variation of this. It was natural to wonder how many weights, and which weights, were really necessary to weigh a given quantity. Bachet asked:

108. What is the least number of weights that can be used on a scale pan to weigh any integral number of pounds from 1 to 40 inclusive, if the weights can be placed in either of the scale pans?

Bachet's *Diophantos* is most famous because it was in the margin of his own copy that Pierre de Fermat wrote a comment on Diophantos's Book II, problem 8, to solve $x^2 + y^2 = a^2$ in integers: 'it is impossible to separate a cube into two cubes, or a biquadrate [fourth powers] into two biquadrates, or in general any power higher than the second into two powers of like degree; I have discovered a truly remarkable proof which this margin is too small to contain.'

This theorem became known as Fermat's Last Theorem, and remains unresolved to this day, though it is widely suspected that it is true.

In his commentary on Diophantos VI, 18, Bachet asked for a triangle with rational sides and a rational altitude, which means that the triangle also has a rational area. Because the area of a triangle can be calculated from the sides using Heron's formula,

$$A = \sqrt{s(s-a)(s-b)(s-c)}$$

where $s = \frac{1}{2}(a + b + c)$, such triangles are called 'Heronian'. Since all the measurements can be multiplied up to make them integers, Heronian triangles are often considered to have integral sides and area.

109. What are the sides and area of the unique Heronian triangle, one of whose altitudes and its three sides are consecutive numbers?

110. What are the three Heronian triangles, which are not right-angled, whose area and perimeter are equal?

111. The area of a Heronian triangle is always a multiple of 6. What is the unique Heronian triangle with area 24?

Henry van Etten

Henry van Etten (1624) was the author of *Mathematical Recreations, Or a Collection of sundrie excellent Problemes out of ancient and moderne Phylosophers Both usefull and Recreative*, published in French in 1624 and first published in English translation in 1633.

It was a compilation, naturally, including questions from the *Greek Anthology* and copying from Bachet, on which it was certainly based, but containing much extra and varied material.

It is also an important early work on conjuring, containing the first description of the 'Inexhaustible Barrel' or 'Any Drink Called For', which allows a variety of drinks to be poured at the magician's whim from the same spout.

Its mathematical problems are mixed up with mechanical puzzles and experiments in optics and hydrostatics, instructions on the making of fireworks, and tips such as 'How to keep wine fresh without ice or snow in the height of summer'.

The first mechanical problem is to break a staff resting on two glasses of water, attributed to Aristotle. The solution is to hit it sufficiently sharply in the middle, and it will break, due to the inertia of the staff.

112. Arrange three knives so that they 'hang in the air without being supported by anything but themselves'.

Variants in Victorian puzzle books demanded how three knives might be used to support a drinking glass, in the ample space between three other drinking glasses placed on the table with more than enough space for a fourth glass to be placed on the table between them.

Several of the following problems also appear two centuries later as popular Victorian amusements.

113. How can a stick be made to balance securely on the tip of a finger?

114. You have a strong staff, and a bucket almost full of water. Required to support the bucket over the edge of the table.

115. How can a bottle be lifted using only a single straw?

116. What shape of bung can be used to plug three different holes, one square, one triangular and one circular?

117. How may a man have his head upwards and his feet upwards at the same time?

118. Two men ascend two ladders, at the same speed, and yet they get further apart. Explain.

119. Where can a man look south in all directions?

120. How can a compass with a fixed opening be used to draw circles of different sizes?

121. How can an oval be drawn with one turn of the compass?

122. Two horses were born at the same time, travelled the world, and then died at the same time, but did not live to the same age. How was this possible?

123. 'Three women, A, B, C, carried apples to a market to sell. A had 20, B, 30, and C, 40; they sold at the same price, the one as the other; and, each having sold all their apples, brought home as much money as each other. How could this be?'

124. Why must there certainly be at least two people in the world with exactly the same number of hairs on their head?

•

Pierre de Fermat (1601–65) was a lawyer by profession and an amateur mathematician of genius who contributed to the development of the calculus and the invention of analytical geometry, and who leapt beyond Diophantos to found the modern theory of numbers.

He posed the following problem to Torricelli, Galileo's famous pupil, who invented the barometer:

125. Find the point whose sum of distances from the vertices of a given triangle is a minimum.

This problem has a natural appeal, because it can be interpreted as asking for the shortest road network that will join three towns at the vertices of the triangle. The next problem occurs first in Urbino d'Aviso's treatise on the sphere (1682):

126. A strip of paper can be transformed into a pentagon. How?

Prince Rupert's Cube   Prince Rupert was a nephew of Charles I of England, a soldier in the Civil War and an inventor and early member of the Royal Society. He enquired:

127. What is the largest cube that can be passed through a square hole cut in a given cube?

Sir Isaac Newton (1642–1727) composed a book on elementary algebra, his *Arithmetica Universalis* (1707), in which this problem occurs:

128.   If a cows graze b fields bare in c days,
            and a' cows graze b' fields bare in c' days,
            and a'' cows graze b'' fields bare in c'' days,

what is the relationship between the nine magnitudes a to c''?

In 1693 Samuel Pepys the diarist and Secretary for the Navy wrote to Newton with this query, a natural question for a gambler:

129. Which is more likely, to throw at least 1 six with 6 dice, at least 2 sixes with 12 dice, or at least 3 sixes with 18 dice?

The misaddressed letters

Niclaus Bernoulli (1687–1759), one of the extraordinary Bernoulli family which produced nine outstanding mathematicians in three generations, considers this problem (but with n letters instead of ten):

130. A correspondent writes ten letters and addresses ten envelopes, one for each letter. In how many ways can all the letters be placed in the wrong envelopes?

131. A related question: if seven letters are placed in seven envelopes randomly, how many letters would you expect, on average, to find in their correct envelopes?

•

Leonhard Euler (1707–83) was one of the most versatile mathematicians of all time, as well as one of the greatest. Here are three of the problems he considered.

132. The Knight's Tour   How can a knight make a complete tour of the chessboard shown on p. 38, visiting each square once and only once, and ending up a knight's move from its starting square – so that the circuit is continuous?

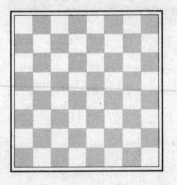

The Bridges of Königsberg   In the town of Königsberg there were
seven bridges across the river Pregel. This popular question was
answered by Euler in 1736:

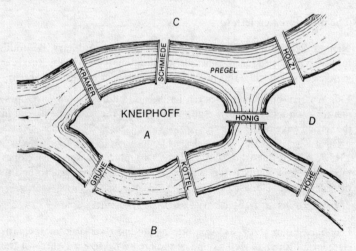

133. Is it possible to go for a walk, crossing each bridge once, but not
crossing any bridge twice?

This was the first ever problem in what is now called *graph theory*. A
graph is a set of points, called vertices or nodes, joined by a set of
lines, called edges. A vertex where an odd number of edges meet is
called an odd vertex, naturally. Graph theory poses many problems,
some of them very simple and simply puzzling:

134. Why is the number of odd vertices in a graph always even?

The Thirty-six Officers Problem Euler considered the problem of placing thirty-six officers, comprising a colonel, lieutenant-colonel, major, captain, lieutenant and sub-lieutenant from each of six regiments, in a square array so that no rank or regiment will be repeated in any row or column.

This problem turns out to be impossible, but the same problem with twenty-five officers is not:

135. How can five each of As, Bs, Cs, Ds and Es be placed in these cells so that no letter is repeated in any row or column?

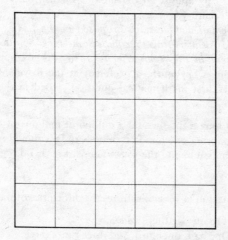

The Ladies' Diary or Woman's Almanac, 1704–1841

The Ladies' Diary was first published in 1704 and consisted initially of recipes, sketches of notable women, and articles on education and health, naturally appealing to its readership.

Within a short time, however, its contents changed, to be replaced by rebuses, enigmas and mathematical questions. That it not only survived but flourished is a blow in the eye to those who suppose that women cannot be interested in mathematics, and proof that caricatures of women as mathematically incapable were less well-established in the early eighteenth century than the late twentieth century. Although men soon proposed and answered many of the questions, women continued to contribute as posers and solvers.

The problems were initially proposed, in the manner of the times, in verse, but, mathematics not lending itself to versification, this practice was soon abandoned.

Although it was not a compilation, some venerable problems were proposed. The first question that is identified as 'Solution by a Lady' (respondents were often anonymous or identified by aliases, such as 'Anne Philomathes') concerned grains of wheat on a chessboard, as payment for sixty-four diamonds: one grain on the first square, two on the second and so on.

In subsequent years, many of the questions were very difficult, and were answered by almost all the famous English mathematicians of the eighteenth century.

The Mathematical Questions Proposed in the Ladies' Diary, and their Original Answers, together with Some New Solutions, from its Commencement in the Year 1704 to 1816 was a compilation, the work of Thomas Leybourn, a professor at the Royal Military College. The very first mathematical question, posed in the year 1707, was:

136. 'In how long a time would a million of millions of money be in counting, supposing one hundred pounds to be counted every minute without intermission, and the year to consist of 365 days, 5 hours, 45 minutes?'

This early question in verse illustrates the difficulties of the form:

137.   If to my age there added be,
            One half, one third, and three times three;
            Six score and ten the sum you'll see,
            Pray find out what my age may be.

138. 'A person remarked that upon his wedding day the proportion of his own age to that of his bride was as 3 to 1; but fifteen years afterwards the proportion of their ages was 2 to 1. What were their ages upon the day of their marriage?'

Question 36 was posed by Mrs Barbara Sidway:

139. 'From a given cone to cut the greatest cylinder possible.'

Question 42 concerns a maypole which breaks, the tip making a mark on the ground. In other words, it is a variant of problem 66 and more than 1000 years old. Question 51 was also old. The solution noted that the problem appeared in Diophantos, Book V.

In contrast the next three problems have a modern feel:

140. What is the least number which will divide by the nine digits without leaving a remainder?

141. 'There came three Dutchmen of my acquaintance to see me, being lately married; they brought their wives with them. The men's names were Hendrick, Claas, and Cornelius; the women's Geertrick, Catriin, and Anna; but I forget the name of each man's wife.

'They told me that they had been at market, to buy hogs; each person bought as many hogs as they gave shillings for each hog; Hendrick bought twenty-three hogs more than Catriin, and Claas bought eleven more than Geertrick; likewise, each man laid out 3 guineas more than his wife. I desire to know the name of each man's wife?' (A guinea was 21 shillings.)

142. 'Being at so large a distance from the dial-plate of a great clock, that I could not distinguish the figures; but as the hour and minute hands were very bright and glaring,' the correspondent noted that they were in a straight line and pointing upwards to the right. It was evening. What was the time?

•

The Vanishing Square Paradox

William Hooper, in his *Rational Recreations* (1774), proposed the first of many vanishing square paradoxes.

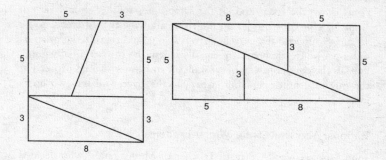

143. The top square has an area $8^2 = 64$. The same four pieces, when reassembled to make the lower figure, form a rectangle $5 \times 13 = 65$. Where has the extra square come from?

144. This is a modern variant. Your task is to reassemble these sixteen pieces to make another 13×13 square, but with an empty square in the centre.

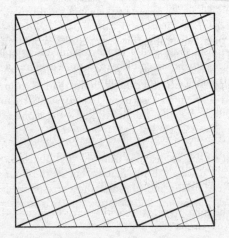

Rowing with and against the Tide

This is another first, which occurs in an arithmetic textbook published in the United States in 1788.

145. 'If, during ebb tide, a wherry should set out from Haverhill, to come down the river, and, at the same time, another should set out from Newburyport, to go up the river, allowing the difference to be 18 miles; suppose the current forwards one and retards the other $1\frac{1}{2}$ miles per hour; the boats are equally laden, the rowers equally good, and, in the common way of working in still water, would proceed at the rate of 4 miles per hour; when, in the river, will the two boats meet?'

Rational Amusements for Winter Evenings

John Jackson was 'A Private Teacher of Mathematics' who decided that there were many puzzles scattered around, but not collected together in one small and convenient volume, so he assembled them himself and wrote *Rational Amusements for Winter Evenings, or, A Collection of above 200 Curious and Interesting Puzzles and Paradoxes relating to Arithmetic, Geometry, Geography, &*, 'Designed Chiefly for Young Persons', which appeared in London in 1821.

From its great rarity it may be inferred that it did not sell many copies, which is a shame because, in addition to its superb title – anticipating the present volume! – it contains many of the classic puzzles and some that had not apparently appeared earlier in print, in particular a collection of ten tree-planting problems and a collection of fifteen variously shaped tiles which resemble a complicated set of Tangram pieces, to be assembled to form a square, a rectangle, a right-angled triangle, a rhombus, and so on.

146. 'It is required to express 100 by four 9s.'

147. 'If from six ye take nine, and from nine ye take ten
(Ye youths, now the mystery explain),
And if fifty from forty be taken, there then,
Shall just half a dozen remain.'

148. 'Place the nine digits, so that the sum of the odd digits may be equal to the sum of the even ones.'

149. 'One third of twelve if you divide,
By just one fifth of seven,
The true result (it has been tried)
Exactly is eleven.'

150. 'Place in a row nine [digits] each different from the others. Multiply them by 8, and the product shall still consist of nine different [digits].'

151. You have 12 pints of wine in a barrel and you wish to divide it into 6 pints for a friend and 6 pints for yourself, but you only have containers holding 7 and 5 pints. How can you succeed?

152. 'With the first nine terms of the geometrical progression 1, 2, 4 . . . to form a product of 4096 each way.' (In other words, the products of the numbers of each vertical column and each horizontal row in the box overleaf must equal 4096.)

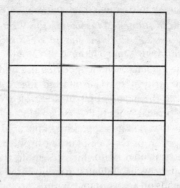

153. 'With the numbers 1, 2, 3, . . . , to 16, to form 34 every way.'

154. 'A Cheshire cheese being put into one of the scales of a false balance, was found to weigh 16 lbs, and when put into the other only 9 lbs. What is the true weight?'

155. 'Mathematicians affirm that of all bodies contained under the same superficies, a sphere is the most capacious; but they have never considered the amazing capaciousness of a body, the name of which is now required, of which it may be truly affirmed that supposing its greatest length 9 inches, greatest breadth 4 inches, and greatest depth 3 inches, yet under these dimensions it contains a solid foot?'

156. 'Divide a circle into four equal parts by three lines of equal length.'

157. 'To make a triangle that shall have three right-angles.'

158. 'To inscribe a square in a given circle, by means of compasses *only*, supposing the centre to be known.'

The next question asks the reader to inscribe a regular dodecagon (12 sides) in a circle under the same conditions.

Jackson's book concludes with no less than sixty 'Geographical Paradoxes', certainly well-calculated to enliven lessons on the globe. One of these depends on acquaintance with a specific physical phenomenon, and is worth relating to illustrate how puzzles in natural philosophy were often mixed in with more logical or mathematical puzzles:

> 'There is a certain place in Great Britain, where, when the tide is in, the sheep may be seen feeding on a certain neighbouring island; yet, when the tide is out, and the water at the lowest, not one can be seen, though they be feeding there at the same instant.'

Jackson's explanation is: 'The place may be the wharf at Greenwich, the Isle of Dogs over against it, and the appearance caused by refraction, when the water is high.'

The remaining puzzles require no such localized knowledge.

159. 'There are three remarkable places on the globe that differ in latitude, as well as in longitude; and yet, all of them lie under the same meridian.'

160. 'There is a particular place on earth, where the winds (although frequently veering round the compass) always blow from the north point.'

161. 'There is a certain village in the Kingdom of Naples, situated in a very low valley, and yet the sun is nearer to the inhabitants thereof, every noon by 3000 miles and upwards, than when he either rises or sets, to those of the said village.'

162. 'There is a certain island, situated between England and France, and yet, that island is farther from France than England is.'

163.   Christians the week's *first* day for sabbath hold,
               The Jews the *seventh*, as they did of old,
               The Turks the *sixth*, as we have oft been told.
               How can these three, in the same place, and day,
               Have each his own true sabbath, tell, I pray.

164. A traveller sets out on a journey, and eventually returns to the place from where he started. During his journey, his head has travelled 12 yards further than his feet, and yet his head remains attached to his body. How is this possible?

From Ozanam to Hutton

Jacques Ozanam was a Frenchman, born in 1640, who wrote a book of recreations based on Bachet and other traditional sources. It was greatly enlarged and improved by Jean Etienne Montucla (born 1725), a friend of Diderot and D'Alembert.

It was translated into English by Charles Hutton, professor at the Royal Military Academy at Woolwich. Edward Riddle's revision of it, called *Recreations in Mathematics and Natural Philosophy* (1840), was the largest collection of mathematical puzzles and recreations published in this country up to that time. The following problems are taken partly from Ozanam's 1741 edition, and partly from Riddle.

165. The hour and minute hands of a watch coincide at noon. When will they once again coincide, during the next 12 hours?

166. 'We are told by Father Sebastian Truchet, of the Royal Academy of Sciences, in a memoir printed [in] 1704 . . . that having seen during the course of a tour which he made to the town of Orleans, some square porcelain tiles, divided by a diagonal into two triangles of different colours . . . he was induced to try in how many different ways they could be joined side by side, in order to form different figures. [Such tiles] form the object of a pastime, called by the French *Jeu de Parquet* . . . a small table, having a border round it, and capable of receiving sixty-four or a hundred small squares . . . with which people amuse themselves in endeavouring to form agreeable combinations.

How many figures can be formed by three squares if the colours of the two halves are black and white and if an edge is placed against a complete edge?

167. If A and B together can complete a task of work in 8 days; and if A and C together take 9 days, and B and C together take 10 days, how much will each man take to do the work by himself?

The next problem is equivalent to the dissection of a Greek Cross into a square:

168. How can five equal squares be dissected and reassembled to form one large square?

169. Demonstrate Pythagoras's Theorem by dissecting the smaller squares to form the larger square, the pieces to be moved by transposition only, without rotation or turning over.

170. Dissect a given rectangle into a square.

171. 'A gentleman wishes to have a silver vessel of a cylindrical form, open at the top, capable of containing a cubic foot of liquor; but being desirous to save the material as much as possible, requests to know the proper dimensions of the vessel.'

172. A man has two wines, one of which sells at 10 shillings per bottle, and the other at 5 shillings. What is the mixture that would sell at 8 shillings a bottle?

173. What is the largest rectangle that can be cut in one piece from this triangular piece of timber?

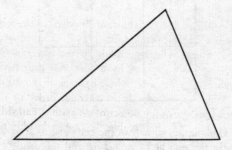

The next problem is attributed to a Mr D—, who said that he got it from M. Buffon, the French naturalist and translator of Newton's *Principia*:

174. Given any irregular polygon, the mid-points of the sides are joined in sequence, and this process is then repeated, again and again. 'It is required to find the point where these divisions will terminate.'

175. 'Given two lines and a point within the angle formed by them, to find the smallest triangle by area that can be cut off.'

176. The Harmonic Square The ancient Greeks considered three important means, the arithmetic and the geometric, which are well known, and the harmonic. The harmonic mean of two numbers is found by taking their reciprocals, finding their average and taking the reciprocal of the result. In other words the harmonic mean of x and y is

$$\frac{1}{\frac{1}{2}\left(\frac{1}{x} + \frac{1}{y}\right)}$$

which simplifies to

$$\frac{2xy}{x + y}$$

With that explanation, how can the cells of this square be filled so that the cells in the middle of each side and the centre cell are each the harmonic means of the numbers sandwiching them? The central number is sandwiched, of course, in four different ways.

177. One player chooses a number less than 11. The second player does likewise and adds his number to the first player's number. The first player again adds a number less than 11, and so on. The player who reaches the grand total of 100 or more is the winner. Is there a winning strategy?

178. The Eight Queens How can eight queens be placed on a chessboard so that no queen attacks any other?

This problem was first posed by Max Bezzel, writing under the pseudonym 'Schachfreund', in the chess magazine *Berliner Schachzeitung*, in 1848. To find all the solutions is extremely difficult, because of the size of the board. An easier problem is:

179. How can 4 (5,6) queens be placed on a 4 × 4 (5 × 5, 6 × 6) board so that no queen attacks any other?

The Victorian Era

Between John Jackson, Riddle's edition of Ozanam, and the end of the century, a wealth of books appeared, with titles such as *The Games Book for Boys and Girls*, *Cassell's Book of Indoor Amusements*, *Card Games and Fireside Fun*, and *The Illustrated Book of Puzzles and Parlour Pastimes: A Repertoire of Acting Charades, Fire-Side Games, Enigmas, Riddles, Charades, Conundrums, Arithmetical and Mechanical Puzzles, Parlour Magic etc*.

They were mostly written for young people, and contained sections of mathematical, mechanical and word puzzles, often a section of magic tricks, simple scientific experiments, plus a wealth of literary puzzles, enigmas, charades, rebuses and maybe chapters on outdoor as well as indoor games and amusements.

The authors showed the usual reliance on old sources, which is why the problems below are not credited to particular books.

According to Dudeney, writing in his *The World's Best Puzzles*, this puzzle has been attributed to Sir Isaac Newton, but Dudeney himself knew of no earlier source than 'a rare book, published in 1821', which was John Jackson's *Rational Amusements*:

180. Tree in a Row How can nine trees be arranged in ten rows with three trees in every row?

Parlour Pastimes posed a variant in verse, with an extra condition:

181.   Ingenious artist, pray dispose,
              Twenty-four trees in twenty-eight rows;
              Three trees I'd have in every row,
              A pond in the midst I'd have also;
              A plan of it I fain would have,
              Which makes me your assistance crave.

The following, on the face of it, is a simpler problem:

182. 'Plant four trees at equal distance from each other.'

183. 'Place twelve counters in six rows in such a manner that there shall be four counters in each row.'

184. You have to divide the number 45 into four parts. To the first part you add 2, from the second part you take 2, the third part you multiply by 2, and the fourth part you divide by 2, so that the *sum* of the addition, the *remainder* of the subtraction, the *product* of the multiplication, and the *quotient* of the division are all equally and precisely the same. How is this possible?

185. 'Having placed eight coins in a row, as under, show how they can be laid or placed in four couples, removing only one at a time, passing over *two* each time.'

                  1   2   3   4   5   6   7   8

186. 'Draw six lines as under, add five other lines, and make the whole form nine.'

(There is also a French version of this puzzle: add three lines to make eight.)

187.   The half of *twelve* is *seven*, as I can show;
              The half of *thirteen* *eight*; can this be so?

188. These dogs are dead, perhaps you'll say;
Add four lines, and then they'll run away.

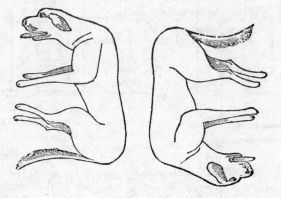

189. 'Of five pieces of wood, or paper, cut in the following shapes, form a cross.'

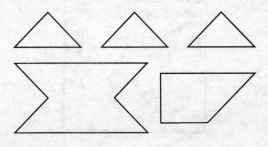

190. From these five shapes, also form a cross:

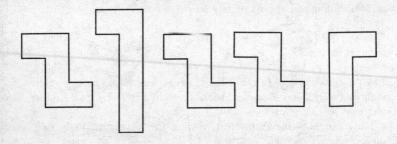

191. 'A charitable individual built a house in one corner of a square plot of ground, and let it to four persons. In the ground were four cherry trees, and it was necessary so to divide it, that each person might have a tree and an equal portion of garden ground. Here is a sketch of the plot. How is it to be divided?'

The next four problems are from *Scientific Amusements* by 'Tom Tit', based on the popular French book *La Science Amusante*.

192. How can an equilateral triangle be constructed by folding a square of paper?

193. How can a ladder be made out of a single sheet of paper, without using gum or other adherent, and this to be effected with three cuts with the scissors?

194. How can a half-crown be passed through a hole the size of a shilling? The old half-crown was approximately 3.1 cm across and the old shilling was approximately 2.3 cm in diameter.

195. What is the largest envelope that can be constructed by folding a rectangle of paper?

196. 'A carpenter had to mend a hole in the floor which was two feet wide and twelve feet long. The board given him to mend it was three feet wide and eight feet long.'

How can he achieve this feat, cutting the board into only two pieces?

197. How can this board, marked as shown, be cut into four identical pieces, so that each piece contains three of the marks, and no mark is cut?

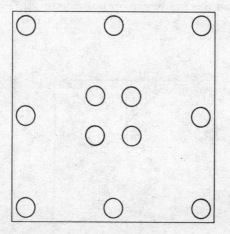

198. Cut a hole in a visiting card large enough for a person to climb through.

199. Place ten coins in a row upon a table. Then, taking up any one of the series, place it upon some other, with this proviso, that you pass over just two coins. Repeat this till there is no single coin left.

200. Arrange the digits 0 to 9 so that they sum to 100.

201. How many animals are concealed in this picture?

202. This is another square with one quarter missing.

How can it be divided between four sons, so that each receives an area identical in shape and size to the others'?

203. How many strokes are necessary to draw this figure, without going over any line twice? A stroke ends as soon as you lift your pencil from the paper.

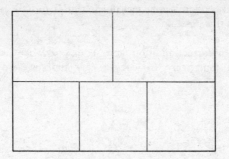

204. A tumbler is resting on three 10p pieces on a table cloth. Under the centre of the tumbler lies a 20p coin. Being thinner than the 10p piece, it can in theory be removed from under the edge of the tumbler without disturbing the tumbler, but you are not allowed to use a knife, or sheet of paper or card or any other suitably thin instrument.

How do you remove the 20p piece?

205. Here are three squares, each composed of four matches. Make them into one by taking one match away, and moving only three others.

206. Here are the same three squares of matches.

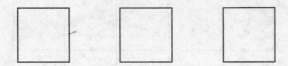

Move three matches to show what matches are made of.

207. Taking one corner of a plain unknotted handkerchief in one hand, and the opposite corner in the other hand, you bring the corners together, and then apart, and Lo and Behold! there is a knot in the handkerchief!

At no time did you release either of the corners of the handkerchief, which remain between the fingers of each hand, exactly as you picked it up. How is this possible?

208. Here is a correct addition sum. Your puzzle is to cover one of the numbers completely, to leave a new addition sum which still totals 1240.

$$
\begin{array}{r}
318 \\
303 \\
300 \\
104 \\
215 \\
\hline
1240
\end{array}
$$

209. **An Easy Solitaire** Each number represents a piece that can jump over any other piece, either vertically, horizontally or diagonally, into an empty square beyond. How can all the pieces be removed, except one, which shall be the 9, which ends up in its original position in the centre?

	1	8	7	
	2	9	6	
	3	4	5	

210. How can four triangles be made with just six matches?

211. These twelve counters are arranged to form six equal squares. Remove just three counters to leave just three equal squares.

212.   Upon a piece of paper draw
              The three designs below;
         I should have said of each shape four,
              Which when cut out will show,
         If joined correctly, that which you
              Are striving to unfold –
         An octagon, familiar to
              My friends both young and old.

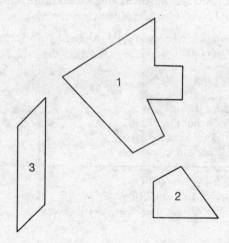

The most comprehensive compilation of this period was *Everybody's Illustrated Book of Puzzles*, selected by Don Lemon (1890), which crammed 794 puzzles, a large majority of them word puzzles, into 125 pages and included most of the puzzles above, and the following. The expression of puzzles in verse was typically Victorian, as was the delight in quibbles and trick questions.

213.   There was a poor man called Johannes Bull,
        Who children did possess, a quiver full;
        And who yet managed somehow to scratch on,
        By the true help of daughter and of son.
        Six little workers had he, each of whom
        Earned something for the household at the loom.
        I will not tell you how much each did gain,
        For I'm a puzzler, and I don't speak plain;
        But, as I would you should possess a clew,
        Some tell-tale facts I'll now disclose to you.
        Week after week, Jane, Ann, Joe, Bet, Rose, Jim,
        Earn ten and tenpence, father says, for him,
        And in this way: The eldest daughter, Jane,
        Gains sevenpence more than sister Anne can gain;
        Ann eightpence more than Joe; while Joe can get
        By his endeavours sixpence more than Bet;
        Bet, not so old, earns not so much as those,
        But by her hands gets fourpence more than Rose;
        Rose, though not up to Jane, yet means to thrive,
        And every week beats Jim by pennies five.

        Now, say what each child worker should receive
        When father draws the cash on pay day eve?

214. Who Can Tell?

        Twice six are eight of us,
        Six are but three of us,
        Nine are but four of us,
           What can we possibly be?
        Would you know more of us?
        I'll tell you more of us.
        Twelve are but six of us,
           Five are but four of us, now do you see?

215.   A row of four figures in value will be
               Above seven thousand nine hundred and three;
               But when they are halved, you'll find very fair
               The sum will be nothing, in truth I declare.

216. *Quibbles*

(a) Add the figure 2 to 191 and make the answer less than 20.

(b) How can I stretch my hands apart, having a coin in each hand, and without bringing my hands together, cause both coins to come into the same hand?

(c) How must I draw a circle round a person placed in the centre of a room so that he will not be able to jump out of it, though his legs should be free?

(d) If five times four are thirty-three, what will the fourth of twenty be?

217. A box has nine ears of corn in it. A squirrel carries out three ears a day, and yet it takes him nine days to carry the corn out. How is this explained?

218. 'A person let his house to several inmates and, having a garden attached to the house, he wished to divide it among them. There were ten trees in the garden and he desired to divide it so that each of the five inmates should have an equal share of the garden and trees. How did he do it?'

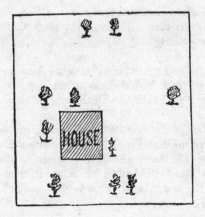

219. This is a trick for a teenager addressing an adult. Let the teenager subtract his or her age from 99, then ask the older person to add this difference to their own age, and then to take the first digit of the amount and add it to the remaining figure.

Query: what will the answer tell the younger person?

The Learned Professor Hoffman

Professor Hoffman's real name was the Reverend Angelo John Lewis. His most famous book, *Puzzles Old and New* (1893), was chiefly devoted to the many popular mechanical puzzles but he also included other Victorian favourites. He also wrote on magic and conjuring.

220. How can this rectangle with two tabs be cut into two pieces to make a complete rectangle?

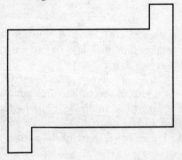

221. 'Required, of the numbers, 1, 2, 3, 4, 5, 6, 7, 8, 9, 0, to compose two fractions, whose sum shall be equal to unity. Each number to be used once, and once only.'

222. 'Required, to find a number of six digits of such a nature that if you transfer the two left-hand digits to the opposite end, the new number thus formed is exactly double the original number.'

223. 'A man goes into a shop and buys a hat, price one guinea. He offers in payment for it a £5 note. The hatter gets the note cashed by a neighbour, the purchaser pocketing his change, £3 19s, and walking off with the hat. No sooner had he left, however, than the neighbour comes in with the news that it is counterfeit, and the hatter has to refund the value.'

How much is the hatter out of pocket by the transaction? (A guinea was 21 shillings.)

224. 'Fifteen matches being laid on the table so as to form five equal squares, required, to remove three matches so as to leave three such squares only.'

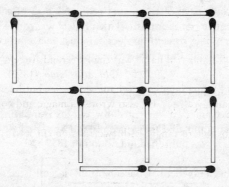

225. How can three matches be taken away to leave a total of seven triangles behind?

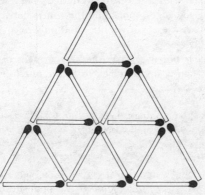

226. An old gentleman was asked who dined with him on Christmas day.

'Well, we were quite a family party,' he replied; 'there was my father's brother-in-law, my brother's father-in-law, my father-in-law's brother-in-law, and my brother-in-law's father-in-law.'

It afterwards transpired that he had dined alone, and yet his statement was correct.

How could this be?

227. What is the difference between six dozen dozen and a half a dozen dozen?

228. Five herrings were divided between five persons. Each had a herring, and yet one remained in the dish. How was this possible?

Mathematicians have often created problems of a popular nature in between their more 'serious' work. Euler was an example. The following examples come from three famous nineteenth-century mathematicians, and one anonymous examination paper.

Hamilton and the Icosian Game

The Icosian Game was invented by W. R. Hamilton, the famous mathematician, and sold to J. Jacques and Son, makers of fine chess sets, for £25. It was published in London in 1859.

229. As 'The Traveller's Dodecahedron' it consisted of a regular dodecahedron, handsomely made in wood, with the names of twenty cities marked at the vertices, in alphabetical order from B for Bruxelles to Z for Zanzibar, with a few letters omitted. They were joined by black lines along the edges, indicating the routes between them, and the object was to visit every town once and only once. How can this be done?

A solid model is not necessary: the arrangement of the towns is indicated in this figure, under which form the puzzle was known as the Icosian Game.

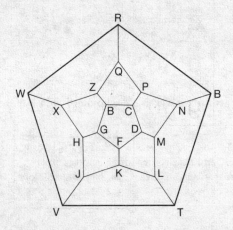

The First Pursuit Problem   The first pursuit problem appeared in a Cambridge Tripos paper of 5 January 1871. It concerned three bugs, each chasing the next. Here is a variation:

230. Four dogs start from the four corners of a square field, of side 100 yards, and race towards each other with constant speed of 3 yards per second, all the dogs starting off in a clockwise direction.

   Where will they meet, and how long will it take them to catch each other?

231. **The Age of Augustus de Morgan**   'Writing in 1864, Professor de Morgan said he was x years old in the year x^2 AD. When was he born?'

Rouse Ball's Mathematical Recreations and Essays

Rouse Ball was the original author of the famous *Mathematical Recreations and Essays*. The first edition was published in 1892, and contained chapters in arithmetical and geometrical recreations, mechanical problems, magic squares and unicursal puzzles, as well as the Essays of the title on the Cambridge Mathematical Tripos, astrology, hyper-space, cryptography and cyphers, and other subjects.

   In subsequent editions, the recreational material became predominant.

232. The figure represents a portion of a chessboard. In the top left corner are eight white pawns, and in the lower right corner, eight black pawns. One move consists of moving a pawn into the adjacent

empty square either horizontally or vertically, or jumping a piece over an adjacent piece, again horizontally or vertically, into the empty square immediately beyond.

No diagonal or backward moves are permitted. How can the black and white pawns be exchanged in the minimum number of moves?

Sylvester and the Postage Stamp Problem

J. J. Sylvester (1814–97) sent the following puzzle to the *Educational Times*, a journal famous in its day for its mathematical problems and the eminent mathematicians who contributed:

233. I have a large number of stamps to the values of 5p and 17p only. What is the largest denomination which I cannot make up with a combination of these two different values?

The Tower of Hanoi and Other Puzzles

Edouard Lucas (1842–91) was a mathematician who studied the Fibonacci sequence and the Lucas sequence, which was named in his honour.

In his *Récréations mathématiques* Lucas discussed Sam Loyd's 'Fifteen' puzzle under the title *Le Jeu du Taquin* and then generalized it to consider any arrangement of squares. Here is a simple case, a rectangular circuit with two additional squares.

234. Four pieces occupy the shaded squares as shown. Is it possible to exchange C for D and also A for B?

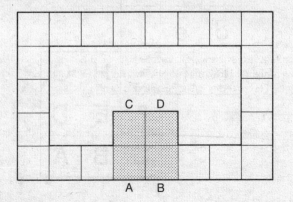

Inspired perhaps by his *Jeu du Taquin*, or merely in a delayed response to the coming of the railways, Lucas posed the first shunting problems:

235. Train A requires to overtake train B, making use of the cul-de-sac line, which is, however, long enough to contain only half of train B. How would you advise the drivers?

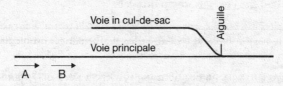

236. A garden is surrounded by a square moat of uniform width. Wishing to cross the moat, to reach the garden, you pick up two planks, each 8 feet long, but the moat is 10 feet wide. Can you cross it safely?

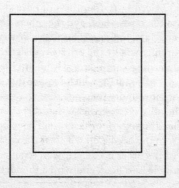

237. Every day at noon a ship leaves Le Havre for New York and another ship leaves New York for Le Havre. The trip lasts seven days and seven nights. How many New York–Le Havre ships will the ship leaving Le Havre today meet during its journey to New York?

238. Lucas's best-known invention is his 'Tower of Hanoi', which was presented to the public in 1883 as the creation of Mr Claus, of the College of Li-Sou-Stian, anagramming Lucas and the Lycée Saint-Louis, where he was then teaching. Hanoi was the capital of Vietnam, an exotic and faraway country that was also a French colony.

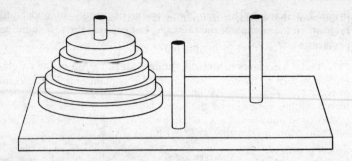

The year after its publication, the following story was published, not by Lucas, to explain the puzzle:

> In the great temple at Benares, says he, beneath the dome which marks the centre of the world, rests a brass plate in which are fixed three diamond needles, each a cubit high and as thick as the body of a bee. On one of these needles, at the creation, God placed sixty-four discs of pure gold, the largest disc resting on the brass plate, and the others getting smaller and smaller up to the top one. This is the Tower of Bramah. Day and night unceasingly the priests transfer the discs from one diamond needle to another according to the fixed and immutable laws of Bramah, which require that the priest on duty must not move more than one disc at a time and that he must place this disc on a needle so that there is no smaller disc below it. When the sixty-four discs shall have been thus transferred from the needle on which at the creation God placed them to one of the other needles, tower, temple, and Brahmins alike will crumble into dust, and with a thunderclap the world will vanish.

The puzzle is to say how many moves are needed to transfer all sixty-four discs.

Lewis Carroll (Reverend Charles Ludwig Dodgson) (1832–98)

Carroll is world-famous as the author of *Alice in Wonderland* and *Alice Through the Looking Glass*, but he was also a witty composer of puzzles and entertainments, as well as being, least importantly to the rest of the world, a lecturer in mathematics at Christ Church, Oxford.

He planned a series of books, never completed, under the general title *Curiosa Mathematica*. The second volume was called *Pillow Problems*, and illustrated his great ability to solve problems in his head. But mathematical and logical problems were only a small part of his output: he invented the word ladder, in which one word is transformed into another, one letter at a time, as BLACK into WHITE, and all his writings were riddled with puns, word-play and logical phantasy.

239. 'A bag contains one counter, known to be either white or black. A white counter is put in, the bag shaken, and a counter drawn out, which proves to be white. What is now the chance of drawing a white counter?'

240. 'If four equilateral triangles be made the sides of a square pyramid: find the ratio which its volume has to that of a tetrahedron made of the triangles.'

241. 'Three points are taken at random on an infinite plane. Find the chance of their being the vertices of an obtuse-angled triangle.'

242. 'I have two clocks: one doesn't go *at all*, and the other loses a minute a day: which would you prefer?'

243. **The Chelsea Pensioners** 'If 70 per cent have lost an eye, 75 per cent an ear, 80 per cent an arm, 85 per cent a leg: what percentage *at least* must have lost all four?'

244. **The Two Omnibuses** Omnibuses start from a certain point, travelling in both directions, every 15 minutes. A traveller, starting on foot along with one of them, meets one coming towards him in $12\frac{1}{2}$ minutes: when will he be overtaken by one?

245. 'Supposing on Tuesday, it is morning in London; in another hour it would be Tuesday morning at the West of England; if the whole world were land we might go on tracing, Tuesday morning, Tuesday morning all the way round, till in twenty-four hours we got to London again. But we *know* that at London twenty-fours hours after Tuesday morning it is Wednesday morning. Where, then, in its passage round the earth, does the day change its name?'

246. 'A rope is supposed to be hung over a wheel fixed to the roof of a building; at one end of the rope a weight is fixed, which exactly counterbalances a monkey which is hanging on to the other end. Suppose that the monkey begins to climb the rope, what will be the result?'

247. 'Put down any number of pounds not more than twelve, any number of shillings under twenty, and any number of pence under twelve. Under the pounds put the number of pence, under the shillings the number of shillings, and under the pence the number of pounds, thus reversing the line. Subtract. Reverse the line again. Add.'

Query: what was Carroll's conclusion?

This next puzzle is related to the river-crossing conundrum which goes back to Alcuin, but the situation has been turned on its side.

248. 'A captive Queen and her son and daughter were shut up in the top room of a very high tower. Outside their window was a pulley with a rope round it, and baskets fastened at each end of the rope of equal weight. They managed to escape with the help of this and a weight they found in the room, quite safely. It would have been dangerous for any of them to come down if they weighed more than 15 lbs more than the contents of the lower basket, for they would do so too quick, and they also managed not to weigh less either. The one basket coming down would naturally of course draw the other up.'

How did they do it?

The Queen weighed 195 lbs, the daughter 165 lbs, the son 90 lbs, and the weight 75 lbs.

This problem is described by Viscount Simon in his memoir of Lewis Carroll:

249. One glass contains 50 spoonfuls of brandy and another glass contains 50 spoonfuls of water. A spoonful of the brandy is transferred to the water, and the mixture is stirred. A spoonful of the mixture is then transferred back to the glass of brandy.

Is there now more brandy in the water, or more water in the brandy?

250. 'Two travellers spend from 3 o'clock till 9 in walking along a level road, up a hill, and home again; their pace on the level being 4 miles an hour, up hill 3, and down hill 6. Find distance walked: also (within half an hour) time of reaching top of hill.'

251. 'A customer bought goods in a shop to the amount of 7s 3d. The only money he had was a half-sovereign, a florin, and a sixpence: so he wanted change. The shopman only had a crown, a shilling, and a penny. But a friend happened to come in, who had a double-florin, a half-crown, a fourpenny bit, and a threepenny bit.

Could they manage it? (A half-sovereign was 10 shillings or 120 pence; a florin was 2 shillings or 24 pence; a crown was 5 shillings or 60 pence.)

252. The Tangram, this dissection of a square into seven pieces from which any number of shapes can be composed, goes back at least as far as the middle of the eighteenth century in China.

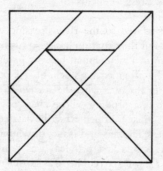

It is no surprise that it appealed to Lewis Carroll. Here are four of the Alice figures as Tangrams. How are they composed?

Loyd and Dudeney

Sam Loyd (1841–1911) and Henry Ernest Dudeney (1857–1930) will always be bracketed as the two greatest puzzle-composers of all time. Not only did their lives largely overlap, in an era when puzzles of all kinds were exceptionally popular and newspapers and magazines were eager to cater for their readers' enthusiasms, but they even worked together for a short period, often shared each other's ideas (or sometimes pinched them – Loyd seems to have done rather more of the pinching) and competed in presenting ingenious new ideas, or familiar ideas in new dress.

Sam Loyd

Sam Loyd was born in Philadelphia, but his parents soon moved to New York, where he attended high school. He considered being an engineer, but gave up the idea when he started to make money from his puzzles. Loyd was a prodigy whose chess problems alone made him famous. He was just fourteen when he started to attend a chess club with his brothers Thomas and Isaac, of whom Isaac also became a noted problemist. Sam's first problem was published in the same year, and by the age of sixteen he was problem editor of *Chess Monthly*, co-edited by Paul Morphy. But by his late teens he had already produced the stunning puzzle of the riderless horses, which the circus owner and showman P. T. Barnum bought from him and sold as 'P. T. Barnum's Trick Donkey'. Loyd had taken the old puzzle of the two dogs (problem 188) and given it a brilliant new twist.

Many years later he produced an even more amazing puzzle, the 'Get off the Earth' paradox. This is how he described the circumstances of its creation, in the *Strand* magazine (January 1908, reprinted in *Sam Loyd and His Chess Problems*, p. 113).

> Unfortunately, it came out in a bad year and did not achieve the success of some of the others. It was developed under rather odd conditions. My son, who thinks I can do anything, said to me one morning, 'Here's a chance, Pop, for you to earn $250,' and he threw a newspaper clipping to me across the breakfast table. It was an offer by Percy Williams of that amount for the best device for advertising Bergen Beach, which he was about to open as a pleasure resort. I said I would take a chance at it, and a few days later I had worked out the Chinaman puzzle. It

*Cut out the three rectangles and rearrange them so
that the two jockeys are riding the two horses.*

consisted of two concentric pieces of cardboard, fastened
together so that the smaller inner one, which was circular,
moved slightly backward and forward, on a pivot, producing
the mystery. As you looked at them, there were thirteen
Chinamen plainly pictured. Move the inner card around a little
and only twelve Chinamen remained. You couldn't tell what
had become of the other Chinaman, try as you would. Scientists
tried it without success, and indeed no single absolutely correct
analysis was ever submitted. Well, on my way to show the
puzzle to Williams, I stopped at the *Brooklyn Eagle* office to ask
Anthony Fiala, their artist and an old friend of mine, to touch it
up a bit for me. I could draw pretty well, but of course he knew
more about it than I did. He was so taken with the puzzle that

he insisted on showing it first to the editor, then to the publisher, and finally to the proprietor of the paper. They all wanted to buy it, but I told them it was disposed of. Finally they proposed that I should run a puzzle department for the *Eagle*; and before I left them they had given me an order for $250 worth of copies of the puzzle, and agreed to a salary of $50 a week for the puzzle column.

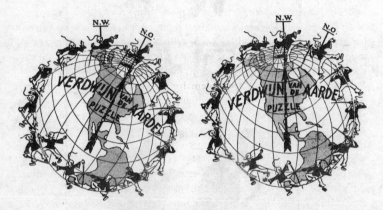

The element of trickery always appealed to Loyd. He once gave a display of mind-reading, using his son as stooge. His son appeared to give correct answer after correct answer as Loyd held up a sequence of cards behind his back, yet his son was only miming – Loyd provided the answers himself by ventriloquism. It was said that a family servant-girl had once left their service because she heard 'voices' every day in the parlour chimney.

Loyd was an excellent mimic, and enjoyed magic tricks and sleight of hand and telling wonderful stories with which he amused his own children – shades of Lewis Carroll.

He was also a self-taught wood-engraver and cartoonist, as well as writer, publisher and editor. He once edited a mechanics trade paper, and produced for a number of years *Sam Loyd's Puzzle Magazine*.

In between the trick donkeys and the vanishing Chinamen, and in complete contrast, he produced in 1878 the '14–15' puzzle (problem 258), which was the Rubik's Cube of the 1870s, and more. The craze swept America, where employers posted notices forbidding employees to play with the puzzle during office hours, and then crossed to Europe. In Germany, Deputies in the Reichstag were observed huddled over the little squares, in France it was called 'le Jeu de Taquin' (see p.

64) and was described as a greater scourge than alcohol or tobacco.

Its spread was aided by the offer of a $1000 prize, a very large sum in those days, for the first person to achieve a particular apparently innocuous position. Loyd had first asked a New York newspaper owner to put up the prize but he refused, and so Loyd offered it himself, risking nothing because the puzzle was impossible – which also meant that Loyd was not able to patent it.

According to Alain C. White, his friend and author of *Sam Loyd and His Chess Problems*, 'Ideas came to him with great fecundity, often too rapidly for him to analyse them completely. Yet his powers for rapid analysis were almost unrivalled. He could see an idea from many sides at once; first always from the point of view of a puzzle, then from the humorous standpoint, finally from the artistic aspect.'

It is a curiosity, and a problem for psychologists to explain, that someone so creative and so brilliant at chess-problem composition should not have been a better over-the-board player or as good a mathematician as Dudeney. Fortunately, in his puzzles his fecund imagination, his ingenuity and his sense of humour were given full reign.

After his death, his son, also named Sam, continued to produce puzzles for newspapers, and also made collections of his father's work. He lacked his father's talents, but possessed ample nerve, exploiting the fact that they shared the same name to write as late as 1928, *Sam Loyd and his Puzzles: An Autobiographical Review*. He had previously compiled, in 1914, *Sam Loyd's Cyclopaedia of 5000 Puzzles, Tricks and Conundrums*. The following puzzles are taken from these two books.

This puzzle perfectly illustrates Loyd's ingenuity and humour, as well as his ability to turn his puzzles into money. The theme is reminiscent of the frog climbing out of the well.

253. 'Many years ago, when Barnum's Circus was of a truth "the greatest show on earth", the famous showman got me to prepare for him a series of prize puzzles for advertising purposes. They became widely known as the Questions of the Sphinx, on account of the large prizes offered to any one who could master them.

'Barnum was particularly pleased with the problem of the cat and dog race, letting it be known far and wide that on a certain day of April he would give the answer and award the prizes, or, as he aptly put it, "let the cat out of the bag, for the benefit of those most concerned".

'The wording of the puzzle was as follows:

'"A trained cat and dog run a race, one hundred feet straight-away and return. The dog leaps three feet at each bound and the cat but two, but then she makes three leaps to his two. Now, under those circumstances, what are the possible outcomes of the race?"

'The fact that the answer was to be made public on the first of April, and the sly reference to "letting the cat out of the bag", were enough to intimate that the great showman had some funny answer up his sleeve.'

Sam Loyd invented the cryptarithm, and it is appropriate that this example should be a long division sum:

254. Can you restore the missing digits?

'The archaeologist is examining a completed problem in long division, engraved on a sandstone boulder. Due to weathering of the rock, most of the figures are no longer legible. Fortunately, the eight legible digits provide enough information to enable you to supply the missing figures.

'It really looks as if there should be scores of correct answers, yet so far as I am aware, only one satisfactory restoration of the problem has been suggested.'

255. How many acres are in the interior triangular lake?

'I went to Lakewood the other day to attend an auction sale of some land, but did not make any purchases on account of a peculiar problem which developed. The land was advertised as shown in the posters on the fence as 560 acres, including a triangular lake. The three plots show the 560 acres without the lake, but since the lake was included in the sale, I, as well as other would-be purchasers, wished to know whether the lake area was really deducted from the land.

'The auctioneer guaranteed 560 acres "more or less". This was not satisfactory to the purchasers, so we left him arguing with katydids, and shouting to the bullfrogs in the lake, which in reality was a swamp.

'The question I ask our puzzlists is to determine how many acres there be in that triangular lake, surrounded as shown by square plots of 370, 116 and 74 acres. The problem is of peculiar interest to those of a mathematical turn, in that it gives a positive and definite answer to a proposition which, according to usual methods, produces one of those ever-decreasing, but never-ending decimal fractions.'

256. Rearrange the six pieces to make the best possible picture of a horse.

The Pony Puzzle

'Many years ago, when I was returning from Europe in company with Andrew G. Curtin, the famous war Governor of Pennsylvania (returning from his post in Russia to seek nomination for president of the United States) we discussed the curious White Horse monument on Uffington Hill, Berkshire, England.

'If you know nothing about that weird relic of the early Saxons, the accompanying sketch will afford an excellent idea of its appearance. It represents the figure of a colossal white horse, several hundred feet long, engraved on the side of the mountain about a thousand feet above the level of the sea and easily seen from a distance of some fifteen miles ... After the white horse had been thoroughly discussed, the Governor banteringly exclaimed, "Now, Loyd, there would be a capital subject for a puzzle."

'Many a good puzzle idea has come from just such a tip. So, with my scissors and a piece of silhouette paper, I speedily improvised the accompanying figure of a horse.

'It would be a simple matter to improve the parts and general form of the old horse, and I did modify it in the version which I afterwards published, but somehow I love the old nag best as first devised, with all its faults, so I now present it as it actually occurred to me.

'The world has been moving rapidly during the last decade, and puzzlers are much sharper than they used to be. In those days very few, probably not one out of a thousand, actually mastered the puzzle, so it will be a capital test of the acumen of the past compared with that of the present generation to see how many clever wits of today can solve it.

'Trace an exact copy of the figure as shown. Cut out the six pieces very carefully, then try to arrange them to make the best possible figure of a horse. That is all there is to it, but the entire world laughed for a year over the many grotesque representations of a horse that can be made with those six pieces.

'I sold over one thousand million [sic] copies of "The Pony Puzzle". This prompts me to say that whereas I have brought out many puzzles, patented numerous inventions, and devoted much time and money, to my sorrow, upon the "big things", more money is made from little things like "The Pony Puzzle", which do not require a five-dollar bill to promote and place on the market.'

257. How would you cut this gingerbread dog's head into two pieces of the same shape?

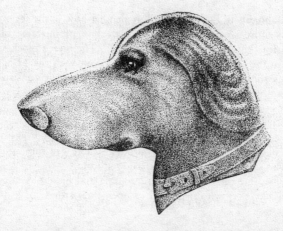

'Here is a practical problem in simple division calculated to baffle some of our puzzlists. You see, Toodles has received the present of a gingerbread dog's head and is told that she must divide it evenly with her little brother. In her anxiety to be fair and equitable in the matter, she wishes to discover some way to divide the cake into two pieces of equal shape and size.

'How many of our clever puzzlists can come to her assistance by showing how the dog's head may be divided?'

258. Slide the numbered blocks into serial order.

'Older inhabitants of Puzzleland will remember how in the seventies I drove the entire world crazy with a little box of movable blocks which became known as the "14–15 Puzzle". The fifteen blocks were arranged in the square box in regular order, but with the 14 and 15 reversed as shown in the above illustration. The puzzle consisted of moving the blocks about, one at a time, to bring them back to the present position in every respect except that the error in the 14 and 15 was corrected.

'A prize of $1000, offered for the first correct solution to the problem, has never been claimed, although there are thousands of persons who say they performed the required feat.

'People became infatuated with the puzzle and ludicrous tales are told of shopkeepers who neglected to open their stores; of a distinguished clergyman who stood under a street lamp all through a wintry night trying to recall the way he had performed the feat. The mysterious feature of the puzzle is that none seem to be able to remember the sequence of moves whereby they feel sure they succeeded in solving the puzzle. Pilots are said to have wrecked their ships, and engineers rush their trains past stations. A famous Baltimore editor tells how he went for his noon lunch and was discovered by his frantic staff long past midnight pushing little pieces of pie around on a plate! Farmers are known to have deserted their plows, and I have taken one such instance as an illustration for the sketch.'

Loyd then gives three further puzzles, of which this is the first:

'Start again with the blocks as shown in the large illustration and move them so as to get the numbers in regular order, but with the vacant square at upper-left-hand corner instead of right-hand corner.

	1	2	3
4	5	6	7
8	9	10	11
12	13	14	15

259. '"What is the age of that boy?" asked the conductor. Flattered by this interest shown in his family affairs, the suburban resident replied:
'"My son is five times as old as my daughter, and my wife is five times as old as the son, and I am twice as old as my wife, whereas grandmother, who is as old as all of us put together, is celebrating her eighty-first birthday today."

'How old was the boy?'

260. An Odd Catch 'Ask your friends if they can write down five odd figures that will add up to fourteen. It is really astonishing how engrossed most people will get, and how much time they will spend over this seemingly simple problem. You must be careful, however, to say "figures" and not "numbers".'

261. Casey's Cow '"Some cows have more sense than the average man," said Farmer Casey. "My old brindle was standing on a bridge the other day, five feet from the middle of the bridge, placidly looking into the water. Suddenly she spied the lightning express, just twice the length of the bridge away from the nearest end of the bridge, coming toward her at a 90-mile an hour clip.

'"Without wasting a moment in idle speculation, the cow made a dash toward the advancing train and saved herself by the narrow margin of one foot. If she had followed the human instinct of running away from the train at the same speed, three inches of her rear would have been caught on the bridge!"

'What is the length of the bridge and the gait of Casey's cow?'

262. The Missing Link 'A farmer had six pieces of chain of five links each, which he wanted made into an endless piece of thirty links. If it costs eight cents to cut a link open and eighteen cents to weld it again, and if a new endless chain could be bought for a dollar and a half, how much would be saved by the cheapest method?'

263. Rearrange the eight pieces to form a perfect chessboard.

'In the history of France is told an amusing story of how the Dauphin saved himself from an impending checkmate, while playing chess with the Duke of Burgundy, by smashing the chessboard into eight pieces over the Duke's head. It is a story often quoted by chess writers to prove that it is not always politic to play to win, and has given rise to a strong line of attack in the game known as the King's gambit.

'The smashing of the chessboard into eight pieces was the feature which always struck my youthful fancy because it might possibly contain the elements of an important problem. The restriction to eight pieces does not give scope for great difficulty or variety, but not feeling at liberty to depart from historical accuracy, I shall give our puzzlists a simple little problem suitable for summer weather. Show how to put the eight pieces together to form a perfect 8 × 8 checkerboard.

'The puzzle is a simple one, given to teach a valuable rule which should be followed in the construction of puzzles of this kind. By giving no two pieces the same shape, other ways of doing the puzzle are prevented, and the feat is much more difficult of accomplishment.'

264. 'An ancient problem, to be found in many old puzzle books, concerns an army fifty miles long. As the army marches forward at a constant rate, a courier starts at the rear of the army, rides forward to deliver a message to the front, then returns to his position at the rear. He arrives back exactly at the time that the army completed an advance of fifty miles. How far altogether did the courier travel?

'If the army were stationary, he could clearly have to travel fifty miles forward and the same distance back. But because the army is advancing, he must go more than fifty miles to the front, and on his return trip he will travel less than fifty miles because the rear of the army is advancing towards him. It is assumed, of course, that the courier always rides at a constant speed.

'A more difficult puzzle is created by the following extension of the theme. A square army, fifty miles long by fifty miles wide, advances fifty miles at a constant rate while a courier starts at the middle of the rear and makes a complete circuit around the army and back to his starting point. The courier's speed is constant, and he completes his circuit just as the army completes its advance. How far does the courier travel?'

265. Tandem Bicycle 'Three men wish to go forty miles on a tandem bicycle that will carry no more than two at a time while the third man is walking. One man, call him A, walks at a rate of one mile in ten minutes, B can walk a mile in fifteen minutes, and C can walk a mile in twenty minutes. The bicycle travels at forty miles an hour regardless of which pair is riding it. What is the shortest time for all three men to make the trip, assuming, of course, that they use the most efficient method of combining walking and cycling?'

266. A Swiss Puzzle 'This pretty Swiss miss is extremely clever at working geometrical cutting puzzles. She has discovered a way of cutting the piece of red wall paper in her right hand into two pieces that will fit together to form the Swiss flag she is holding in her left hand. The white cross in the centre of the flag is actually a hole in the paper. The cutting must follow the lines ruled on the paper.

'For a second puzzle, the Swiss girl asks you to cut the flag in her left hand into two pieces that will fit together to make a rectangle of five-by-six units.

'Someone once asked the Swiss girl how to make a Maltese cross and she replied, "Pull its tail!"'

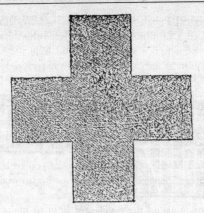

267. Dissecting a Cross 'It is a remarkable fact that a mysterious affinity, or relationship, can be shown to exist between all the ancient signs and symbols, in that each one can be converted into another by rearrangement of its dissected parts. Thus, a Swastika can be changed into a square, the square into a cross, the cross into a triangle, etc., etc.

'One of these interesting transformations consists in dissecting the Greek Cross, shown above, into four similar parts which may be regrouped to present a square, with a central open space in the form of a small Greek Cross. Work out the problem mentally before applying your scissors.'

268. The Trapezoid Puzzle How can these five pieces be variously assembled to form a square, a Greek Cross, a rhombus, a rectangle, and a triangle? All five pieces must be used for each assembly.

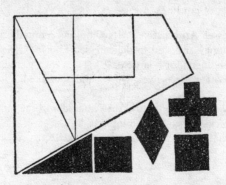

269. A Paradoxical Puzzle This puzzle perfectly illustrates Loyd's ability to spot, in a familiar situation, what others had missed. This diagram shows how an 8 × 8 square can be dissected and reassembled to, apparently, gain an extra square, since the rectangle is 5 × 13.

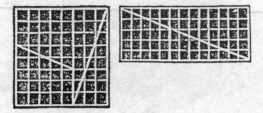

Loyd asks for another way to reassemble the same four pieces which will lose a square.

270. Honest John 'Here is an extension of the "measuring" idea, which you will find more elusive. I first published it in 1900, as follows:

'Honest John, the milkman, says that what he doesn't know about milk is scarcely worth mentioning, but he was nearly flabbergasted the other day when he got out on his route with his two 10-gallon cans full of the lacteal fluid, but minus his measuring cans. Then along came two customers, one with a five-quart pail and the other with a four-quart pail, and they each demanded two quarts of milk in a hurry. In filling the orders John proved himself considerable of a puzzler.

'To measure exactly two quarts of milk in each of those pails is a measuring problem pure and simple, devoid of trick or device, but it requires considerable cleverness to achieve the desired result with the fewest number of pourings.'

271. Delicatessen Arithmetic 'Mrs Simpkins counted out the correct amount of money and said to delicatessen Louis: "Give me a pound and a half of bologna for boarders."

'Louis cut off a piece, weighed it, and remarked: "It weighs 10c over."

'"Then give me half of it, and the remainder of the money will buy 5c worth of pickles," said Mrs Simpkins.

'How much did she expend on the bologna?'

272. Carving a Doughnut 'The design shows the sort of doughnut that buddies claim the Salvation Army lassies turned out "over there". Whether intended or not, this particular doughnut makes an interesting puzzle. Just draw a straight line across the doughnut to show how many pieces you could produce with one straight cut.'

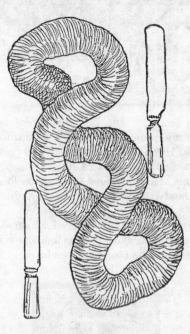

273. Popping the Question 'Danny went over to urge Kate to name the day.

'"This is entirely unexpected," gasped the maiden; "but I will marry you when the week after next is the week before last."

'"Had I received this promise yesterday," said Danny, "the waiting would have been six days shorter."

'Can you tell on what day of the week Danny popped the question?'

274. Make it Square 'This design [overleaf] contains exactly sixty-four little squares, and the puzzle consists in showing how it may be cut into the least possible number of pieces to make a large eight by eight square, with the pattern preserved. How do you do it?'

275. Everything Free 'A little girl visited the food show and ate seventeen different kinds of breakfast food and gathered 10 pounds of sample packages. Then she stepped on the free weighing machine and found that her weight had increased 10 per cent, whereas if she had eaten twice as much breakfast food the gain would have been 11 per cent. What was her weight when she arrived at the food show?'

A Revolutionary Rebus Loyd published hundreds of rebuses and other simple puzzles. This example is included by way of illustration, because of its ingenuity. As Loyd presented it:

276. 'One of the incidents leading to the Revolutionary War, especially interesting to young students, is represented by that monogram. What was the historical event?'

The Longest Queen's Tour This puzzle appeared in *Le Sphinx* in March 1867, and was described by Loyd as the best of its kind he had ever produced and as 'the most difficult puzzle extant':

277. The problem is simply to 'Place the Queen on [a chessboard] and pass her over the entire sixty-four squares and back again to point of beginning in fourteen moves.'

278. The Same Again, Almost 'Pass the Queen over the centre points of all the squares in fourteen straight moves, returning to the starting point.'

('Straight moves' are not limited to the ordinary Queen moves in chess. Any move in a straight line will do.)

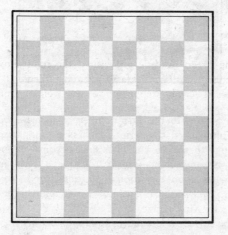

279. Dissecting the Chessboard 'As interesting as any of the dissection puzzles based on the chessboard, is the one which asks:

'What is the maximum number of pieces into which a chessboard can be divided without any two of the pieces being exactly alike?

'Of course, the first step is to mark off a single white square and a single black one. These, alike in size, are dissimilar in colour. There we have the idea. Two pieces may be alike in form and size yet dissimilar in number or arrangement of white and black squares. Pieces are considered unalike if dissimilar in any respect. It is a pretty problem and not too difficult.'

280. Knight Dissection 'Divide the chessboard [overleaf], on the lines, into four exactly equal parts, so that there shall be one of the Knights in each of the parts.'

281. From the Start. In the first number of the *American Chess Journal* Loyd introduced the series of chess puzzles based on the ordinary line-up of the pieces which has since become so famous.

'It will not be amiss,' he wrote, 'to have a little impromptu exhibition, bearing upon conditional positions produced from the position of the forces as arranged for actual play. I find two by Breitenfeld, one by Max Lange, some from "Sissa", Dr Moore, etc., but as all can be solved in less moves than intended by the authors, I give them under one heading, without authorship, and I have thrown in a few similar ideas that occurred to me, elucidated in a sketch.'

(a) If both parties move the same moves, how can the first player mate in four moves?

(b) If both parties make the same moves, how can the first player self-mate on the eighth move?

(c) Find how discovered checkmate can be effected in four moves.

(d) Find how a stalemate might result in ten moves.

(e) Find a game wherein perpetual check can be forced from the third move.

Henry Dudeney

Henry Dudeney was born in the village of Mayfield in Sussex. His paternal grandfather was a self-taught mathematician and astronomer who started as a shepherd and raised himself to the position of schoolmaster in the town of Lewes. Dudeney's father was also a schoolmaster, but Dudeney himself did not go to college and was also a self-taught mathematician.

He enjoyed games and was a good chess player, though, like Sam Loyd, he was a better problemist, as might be expected. He also played croquet, a game that might have been designed for puzzlists, and entertained children with displays of magic and legerdemain.

He started composing puzzles under the pseudonym 'Sphinx', and for a while he collaborated with Sam Loyd. When their collaboration ended, Dudeney published under his own name in a variety of magazines: the *Strand*, *Cassell's*, the *Queen*, *Tit-Bits*, the *Weekly Dispatch* and *Blighty*. So, like modern television stand-up comedy writers, he had to keep up a constant flow of ideas. Sam Loyd was in the same position. This makes it all the more astonishing that their levels were so consistently high. Loyd showed greater ingenuity in exploiting his puzzles, especially for advertising. He had an uncanny knack for appealing to the public. Yet Dudeney was much the better mathematician, and his puzzles are more mathematically sophisticated, without requiring any mathematics beyond the most elementary.*

Dudeney was interested in the psychology of puzzles and puzzle-solving. In the original preface to *A Puzzle-Mine* he asserted that 'The fact is that our lives are largely spent in solving puzzles; for what is a puzzle but a perplexing question? And from our childhood upwards we are perpetually asking questions or trying to answer them.'

---

* I am indebted to Martin Gardner's Introduction to *536 Puzzles and Curious Problems* for some of this information.

But he was also a man of his age. In the same preface he remarks that 'The solving of puzzles consists merely in the employment of our reasoning faculties, and our mental hospitals are built expressly for those unfortunate people who cannot solve puzzles.' Elsewhere he remarks that 'The history of [mathematical puzzles] entails nothing short of the actual story of the beginnings and development of exact thinking in man.'

Half a century later we are more aware of the roles of insight and imagination, and the 'Aha!' response, which are more than the exercise of logic or reason. Dudeney also supposed that puzzles had great value in training the mind, a natural assumption in the days when many educational theorists still believed in the idea of mental training. We would put that rather differently: it is mathematics teachers today who most exploit puzzles and mathematical recreations to entice their pupils and to illuminate mathematical ideas.

Fortunately, just as an artist may have a naïve theory of their own art, so Dudeney's puzzles are not limited by his own interpretations of them. They exhibit a wealth of imagination and ingenuity, even artistry . . .

The following puzzles are selected from *The Canterbury Puzzles* (1907), *Amusements in Mathematics* (1917) and *Modern Puzzles* (1926). The title puzzles of *The Canterbury Puzzles* are a sequence of problems proposed by 'A chance-gathered company of pilgrims, on their way to the shrine of Saint Thomas à Becket at Canterbury, met at the Tabard Inn, later called the Talbot, in Southwark [whose] host proposed that they should beguile the ride by each telling a tale to his fellow-pilgrims.'

282. The Haberdasher's Puzzle 'Many attempts were made to induce the Haberdasher, who was of the party, to propound a puzzle of some kind, but for a long time without success. At last, at one of the Pilgrim's stopping-places, he said that he would show them something that would "put their brains into a twist like unto a bell-rope". As a matter of fact, he was really playing off a practical joke on the company, for he was quite ignorant of any answer to the puzzle that he set them. He produced a piece of cloth in the shape of a perfect equilateral triangle, as shown in the illustration, and said, "Be there any among ye full wise in the true cutting of cloth? I trow not. Every man to his trade, and the scholar may learn from the varlet and the wise man from the fool. Show me, then, if ye can, in what manner this piece of cloth may be cut into four several pieces

that may be put together to make a perfect square."

'Now some of the more learned of the company found a way of doing it in five pieces, but not in four. But when they pressed the Haberdasher for the correct answer he was forced to admit, after much beating about the bush, that he knew no way of doing it in any number of pieces. "By Saint Francis," saith he, "any knave can make a riddle methinks, but it is for them that may to rede it right." For this he narrowly escaped a sound beating. But the curious point of the puzzle is that I have found that the feat really may be performed in so few as four pieces, and without turning over any piece when placing them together. The method of doing this is subtle, but I think the reader will find the problem a most interesting one.'

283. The Spider and the Fly 'Inside a rectangular room, measuring 30 feet in length and 12 feet in width and height, a spider is at a point in the middle of one of the end walls, 1 foot from the ceiling, as at A; and a fly is on the opposite wall, 1 foot from the floor in the centre,

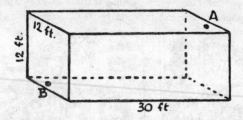

as shown at B. What is the shortest distance that the spider must crawl in order to reach the fly, which remains stationary? Of course the spider never drops or uses its web, but crawls fairly.'

284. Catching the Hogs 'In the illustration Hendrick and Katrun are seen engaged in the exhilarating sport of attempting the capture of a couple of hogs.

'Why did they fail?

'Strange as it may seem, a complete answer is afforded in the little puzzle game that I will now explain.'

[Dudeney instructs the reader to represent the Dutchman and his wife, and the two hogs, by four counters, on squared paper.]

'The first player moves the Dutchman and his wife one square each in any direction (but not diagonally), and then the second player moves both pigs one square each (not diagonally); and so on, in turns, until Hendrick catches one hog and Katrun the other.

'This you will find would be absurdly easy if the hogs moved first, but this is just what Dutch pigs will never do.'

285. Making a Flag 'A good dissection puzzle in so few as two pieces is rather a rarity, so perhaps readers will be interested in the following. The diagram represents a piece of bunting, and it is required to cut it into two pieces (without any waste) that will fit together and form a perfectly square flag, with the four roses symmetrically placed. This would be easy enough if it were not for the four roses, as we should merely have to cut from A to B and insert the piece at the bottom of the flag. But we are not allowed to cut through any of the roses, and therein lies the difficulty of the puzzle.'

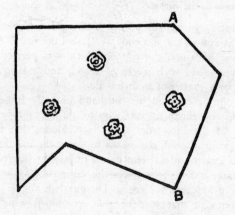

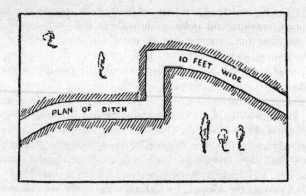

286. Bridging the Ditch As Dudeney describes it, this plan is of a ditch which is 10 feet wide, and filled with water. The King's Jester desperately requires to cross the ditch, but he cannot swim. The only equipment he can find is a heap of eight planks, each of which is no more than 9 feet long.

How can he cross the ditch to safety?

Lady Isabel's Casket This puzzle is the first appearance of the idea of a 'squared square', that is, a square dissected into distinct smaller squares, though Dudeney has to resort to a narrow rectangular strip to fill a portion of the surface.

287. 'Sir Hugh's young kinswoman and ward, Lady Isabel de Fitz-arnulph, was known far and wide as "Isabel the Fair". Amongst her treasures was a casket, the top of which was perfectly square in shape. It was inlaid with pieces of wood, and a strip of gold ten inches long by a quarter of an inch wide.

'When young men sued for the hand of Lady Isabel, Sir Hugh promised his consent to the one who would tell him the dimensions of the top of the box from these facts alone: that there was a rectangular strip of gold, ten inches by 1/4-inch; and the rest of the surface was exactly inlaid with pieces of wood, each piece being a perfect square, and no two pieces the same size. Many young men failed, but one at length succeeded. The puzzle is not an easy one, but the dimensions of that strip of gold, combined with those other conditions, absolutely determine the size of the top of the casket.'

288. The Fly and the Cars   A road is 300 miles long. A car, A, starts at noon from one end and goes throughout at 50 miles an hour, and at the same time another car, B, going uniformly at 100 miles an hour, starts from the other end together with a fly travelling at 150 miles an hour. When the fly meets car A, it immediately turns and flies towards B.

(1) When does the fly meet B?

The fly then turns towards A and continues flying backwards and forwards between A and B.

(2) When will the fly be crushed between the cars if they collide and it does not get out of the way?

289. Crossing the Moat   'I [the King's Jester, still adventuring] was now face to face with the castle moat, which was, indeed, very wide and very deep. Alas! I could not swim, and my chance of escape seemed of a truth hopeless, as, doubtless, it would have been had I not espied a boat tied to a wall by a rope. But after I had got into it I did find that the oars had been taken away, and that there was nothing that I could use to row me across. When I had untied the rope and pushed off upon the water the boat lay quite still, there being no stream or current to help me. How, then, did I yet take the boat across the moat?'

290. The Crescent and the Cross   'When Sir Hugh's kinsman, Sir John de Collingham, came back from the Holy Land, he brought

with him a flag bearing the sign of a crescent, as shown in the illustration. It was noticed that de Fortibus spent much time in examining this crescent and comparing it with the cross borne by the Crusaders on their own banners . . . [He] explained that the crescent in one banner might be cut into pieces that would exactly form the perfect cross in the other. It is certainly rather curious; and I shall show how the conversion from crescent to cross may be made in ten pieces, using every part of the crescent. The flag was alike on both sides, so pieces may be turned over where required.'

291. The Riddle of St Edmondsbury ' "It used to be told at St Edmondsbury," said Father Peter on one occasion, "that many years ago they were so overrun with mice that the good abbot gave orders that all the cats from the country round should be obtained to exterminate the vermin. A record was kept, and at the end of the year it was found that every cat had killed an equal number of mice, and the total was exactly 1,111,111 mice. How many cats do you suppose there were?'

292. The Cigar Puzzle 'Two men are seated at a square-topped table. One places an ordinary cigar (flat at one end, pointed at the other) on the table, then the other does the same, and so on alternately, a condition being that no cigar shall touch another. Which player should succeed in placing the last cigar, assuming that they each will play in the best possible manner? The size of the table top and the size of the cigar are not given, but in order to exclude the ridiculous answer that the table might be so diminutive as only to take one cigar, we will say that the table must not be less than 2 feet square and the cigar not more than $4\frac{1}{2}$ inches long. With those restrictions you may take any dimensions you like. Of course we assume that all the cigars are exactly alike in every respect. Should the first player, or the second player, win?'

293. The Damaged Measure 'Here is a new puzzle that is interesting, and it reminds me, though it is really very different, of the classical problem by Bachet concerning the weight that was broken in pieces which would then allow of any weight in pounds being determined from one pound up to a total weight of all the pieces. In the present case a man has a yard-stick from which 3 inches have been broken off, so that it is only 33 inches in length.

'Some of the graduation marks are also obliterated, so that only eight of these marks are legible; yet he is able to measure any given number of inches from 1 inch up to 33 inches. Where are these marks placed?'

294. Exploring the Desert 'Nine travellers, each possessing a motor-car, meet on the eastern edge of a desert. They wish to explore the interior, always going due west. Each car can travel forty miles on the contents of the engine tank, which holds a gallon of petrol, and each can carry nine extra gallon tins of petrol and no more. Unopened tins can alone be transferred from car to car. What is the greatest distance to which they can enter the desert without making any depots for petrol for the return journey?'

295. A Puzzle with Pawns 'Place two pawns in the middle of the chessboard, one at Q4 and the other at K5. Now, place the remaining fourteen pawns (sixteen in all), so that no three shall be in a straight line in any possible direction.

'Note that I purposely do not say queens, because by the words "any possible direction" I go beyond attacks on diagonals. The pawns must be regarded as mere points in space – at the centres of the squares.'

296. The Game of Bandy-ball Bandy-ball, cambuc, or goff (the game so well known today by the name of golf), is of great antiquity, and was a special favourite at Solvemhall Castle. Sir Hugh de Fortibus was himself a master of the game, and he once proposed this question.

'They had nine holes, 300, 250, 200, 325, 275, 350, 225, 375, and 400 yards apart. If a man could always strike the ball in a perfectly straight line and send it exactly one of two distances, so that it would either go towards the hole, pass over it, or drop into it, what would the two distances be that would carry him in the least number of strokes round the whole course?

'Two very good distances are 125 and 75, which carry you round in twenty-eight strokes, but this is not the correct answer. Can the reader get round in fewer strokes with two other distances?'

297. The Noble Demoiselle 'Seated one night in the hall of the castle, Sir Hugh desired the company to fill their cups and listen while he told the tale of his adventure as a youth in rescuing from captivity a noble demoiselle who was languishing in the dungeon of a castle belonging to his father's greatest enemy ... Sir Hugh produced a plan of the thirty-five cells in the dungeon and asked his companions to discover the particular cell that the demoiselle occupied. He said that if you started at one of the outside cells and passed through every doorway once, and once only, you were bound to end at the cell that was sought. Can you find the cell? Unless you start at the correct outside cell it is impossible to pass through all the doorways once and once only.'

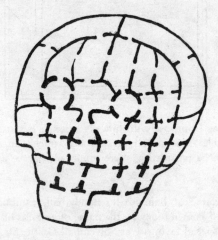

298. The Trusses of Hay 'Farmer Tomkins had five trusses of hay, which he told his man Hodge to weigh before delivering them to a customer. The stupid fellow weighed them two at a time in all possible ways, and informed his master that the weights in pounds were 110, 112, 113, 114, 115, 116, 117, 118, 120 and 121. Now, how was Farmer Tompkins to find out from these figures how much every one of the five trusses weighed singly? The reader may at first think that he ought to be told "which pair is which pair" or something of that sort, but it is quite unnecessary. Can you give the five correct weights?'

299. Another Joiner's Problem 'A joiner had two pieces of wood of the shapes and relative proportions shown in the diagram. He wished to cut them into as few pieces as possible so that they could be fitted together, without waste, to form a perfectly square table-top. How should he have done it? There is no necessity to give measurements, for if the smaller piece (which is half a square) be made a little too large or small, it will not affect the method of solution.'

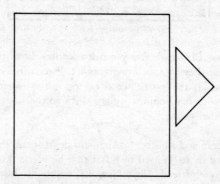

300. The Rook's Tour 'The puzzle is to move the single rook over the whole board, so that it shall visit every square of the board once, and only once, and end its tour on the square on which it starts. You have to do this in as few moves as possible.'

301. The Five Pennies 'Every reader knows how to place four pennies so that they all touch one another. Place three in the form of a triangle, and lay the fourth penny on top in the centre. Now try to do the same with five pennies – place them so that every penny shall touch every other penny.'

302. De Morgan and Another 'Augustus de Morgan, the mathematician, who died in 1871, used to boast that he was x years old in the year x^2. My living friend, Jasper Jenkins, wishing to improve on this, tells me that he was $a^2 + b^2$ in $a^4 + b^4$; that he was $2m$ in the year $2m^2$; and that he was $3n$ years old in the year $3n^4$. Can you give the years in which De Morgan and Jenkins respectively, were born?'

303. Sum Equals Product '"This is a curious thing," a man said to me. "There are two numbers whose sum equals their product. They are 2 and 2, for if you add them or multiply them, the result is 4." Then he tripped badly, for he added, "These are, I find, the only two numbers that have this property."

'I asked him to write down any number, as large as he liked, and I would immediately give him another that would give a like result by addition or multiplication.'

What was Dudeney's method of doing this?

Water, Gas and Electricity This puzzle illustrates the difficulty of deciding which puzzles were invented by Dudeney and Loyd and which were borrowed from each other, or from older sources. Loyd claimed that he invented this puzzle about 1903, Dudeney thinks otherwise. I am inclined to think that such a puzzle – of obvious practical relevance – *ought to be old*, but if so then it is surprising that it does not appear in the commonest Victorian puzzle books.

304. 'There are some half-dozen puzzles, as old as the hills, that are perpetually cropping up, and there is hardly a month in the year that does not bring inquiries as to their solution. Occasionally one of these, that one had thought was an extinct volcano, bursts into eruption in a surprising manner. I have received an extraordinary number of letters respecting the ancient puzzle that I have called "Water, Gas and Electricity". It is much older than electric lighting, or even gas, but the new dress brings it up to date. The puzzle is to lay on water, gas, and electricity, from W, G and E, to each of the three houses, A, B and C, without any pipe crossing another.'

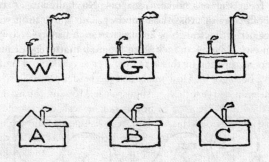

305. The Six Pennies Lay six pennies on the table, and then arrange them as shown overleaf, so that a seventh would fit exactly into the central space. You are not allowed the use of a ruler or any other measuring device, just the six pennies.

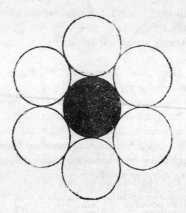

306. Placing Halfpennies 'Here is an interesting little puzzle suggested to me by Mr W. T. Whyte. Mark off on a sheet of paper a rectangular space 5 inches by 3 inches, and then find the greatest number of halfpennies that can be placed within the enclosure under the following conditions. A halfpenny is exactly an inch in diameter. Place your first halfpenny where you like, then place your second coin at exactly the distance of an inch from the first, the third an inch distance from the second, and so on. No halfpenny may touch another halfpenny or cross the boundary. Our illustration will make the matter perfectly clear. No. 2 coin is an inch from No. 1; No. 3 an inch from No. 2; No. 4 an inch from No. 3; but after No. 10 is placed we can go no further in this attempt. Yet several more halfpennies might have been got in. How many can the reader place?'

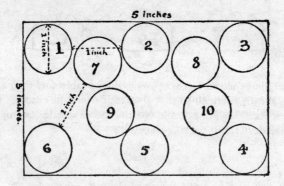

307. The Bun Puzzle   'The three circles represent three buns, and it is simply required to show how these may be equally divided among four boys. The buns must be regarded as of equal thickness throughout and of equal thickness to each other. Of course, they must be cut into as few pieces as possible. To simplify it I will state the rather surprising fact that only five pieces are necessary, from which it will be seen that one boy gets his share in two pieces and the other three receive theirs in a single piece. I am aware that this statement "gives away" the puzzle, but it should not destroy its interest to those who like to discover the "reason why".'

308. The Cardboard Chain   'Can you cut this chain out of a piece of cardboard without any join whatsoever? Every link is solid, without its having been split and afterwards joined at any place. It is an interesting old puzzle that I learnt as a child, but I have no knowledge as to its inventor.'

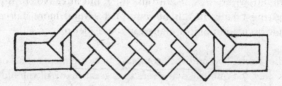

309. The Two Horseshoes   'Why horseshoes should be considered "lucky" is one of those things which no man can understand. It is a very old superstition, and John Aubrey (1626–1700) says, "Most houses at the West End of London have a horseshoe on the threshold." In Monmouth Street there were seventeen in 1813 and seven so late as 1855. Even Lord Nelson had one nailed to the mast of the ship *Victory*. Today we find it more conducive to "good luck" to see that they are securely nailed on the feet of the horse we are about to drive.

'Nevertheless, so far as the horseshoe, like the Swastika and other

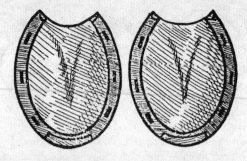

emblems that I have had occasion at times to deal with, has served to symbolize health, prosperity, and goodwill towards men, we may well treat it with a certain amount of respectful interest. May there not, moreover, be some esoteric or lost mathematical mystery concealed in the form of a horseshoe? I have been looking into this matter, and I wish to draw my readers' attention to the very remarkable fact that the pair of horseshoes shown in my illustration are related in a striking and beautiful manner to the circle, which is the symbol of eternity. I present this fact in the form of a simple problem, so that it may be seen how subtly this relation has been concealed for ages and ages. My readers will, I know, be pleased when they find the key to the mystery.

'Cut out the two horseshoes carefully round the outline and then cut them into four pieces, all different in shape, that will fit together and form a perfect circle. Each shoe must be cut into two pieces and all the part of the horse's hoof contained within the outline is to be used and regarded as part of the area.'

310. The Table-Top and the Stools   'I have frequently had occasion to show that the published answers to a great many of the oldest and most widely known puzzles are either quite incorrect or capable of improvement. I propose to consider the old poser of the table-top and stools that most of my readers have probably seen in some form or another in books compiled for the recreation of childhood.

'The story is told that an economical and ingenious schoolmaster once wished to convert a circular table-top, for which he had no use, into seats for two oval stools, each with a hand-hole in the centre. He instructed the carpenter to make the cuts as in the illustration and then join the eight pieces together in the manner shown. So impressed was he with the ingenuity of his performance that he set the puzzle to his geometry class as a little study in dissection. But the remainder of

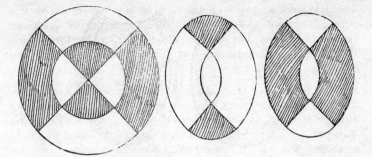

the story has never been published, because, so it is said, it was a characteristic of the principals of academies that they would never admit that they could err. I get my information from a descendant of the original boy who had most reason to be interested in the matter.

'The clever youth suggested modestly to the master that the hand-holes were too big, and that a small boy might perhaps fall through them. He therefore proposed another way of making the cuts that would get over this objection. For his impertinence he received such severe chastisement that he became convinced that the larger the hand-hole in the stools the more comfortable might they be.

'Now what was the method the boy proposed?

'Can you show how the circular table-top may be cut into eight pieces that will fit together and form two oval seats for stools (each of exactly the same size and shape) and each having similar hand-holes of smaller dimensions than in the case shown above? Of course, all the wood must be used.'

Send More Money Loyd was the first inventor of the cryptarithm, in which some or all of the digits in a sum are deleted, and the sum has to be reconstructed, but Dudeney first replaced the missing digits with letters to make a meaningful message – he called it Verbal Arithmetic – a rare example of Dudeney hitting upon a popular point that Loyd missed.

$$\begin{array}{r} SEND \\ MORE \\ \hline MONEY \end{array}$$

311. This is a correct addition sum, in which each different letter stands for a different digit, zero possibly included. What is the original sum?

The Eight Spiders   Just as Dudeney and Loyd used many old puzzles, often adding new ideas of their own, so their puzzles have been exploited by others. The problem of the spider and the fly (problem 283 above) has been especially fruitful in variations. This one is due to Maurice Kraitchik.

312. 'An honourable family of spiders, consisting of a wise mother and eight husky youngsters, were perched on the wall at one end of a rectangular room. Food being scarce, owing to the Second World War, they were grumbling, when an enormous fly landed unnoticed on the opposite wall. If Euclid could have been summoned from his grave (location, alas, unknown), he would have been able to show that both the hunters and the prey were in the vertical plane bisecting the two opposite walls, the spiders eighty inches above the centre and the fly eighty inches below.

'Suddenly one young spider shouted with glee. "Mamma! Look! There's a fly! Let's catch him and eat him!"

'"There are four ways to reach the fly. Which shall we take?" came the eager query from another.

'"You have forgotten your Euclid, my darling. There are *eight* ways to reach the fly. Each of you take a different path, without using any other means of conveyance than your God-given legs. Whoever reaches the goal first shall be rewarded with the largest portion of the prey."

'At the signal given by the mother the eight spiders shot out in eight different directions at a speed of 0.65 mile per hour. At the end of $\frac{625}{11}$ seconds they simultaneously converged on the fly, but found no need of attacking it since its heart had given way at the sight of enemies on all sides.

'What are the dimensions of the room?'

•

313. **The Magic Hexagon**   How can the numbers 1 to 19 be placed in the cells of this hexagon so that all fifteen sums, five each in each of three directions, are equal?

In comparison to normal magic squares, which sum in two directions plus, more often than not, along some diagonals, it might be expected that there would be fewer magic hexagons, summing in three directions. Yet it is surprising that there is just one unique solution to this problem, apart from reflections and rotations, and no solutions at all for any other size of hexagon.

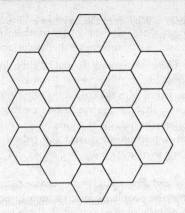

314. The Boat in the Bath Tommy was floating a boat in a tub of water. The boat was initially loaded with a small metal cannon, but then the cannon fell into the water and sank to the bottom, leaving the boat floating as before. No water got into the boat while this happened.

Did the level of the water rise, fall, or stay the same, as a result of the cannon falling overboard?

Arithmorems This is the name given to a simple but elegant type of puzzle, reminiscent of the 'Nuns Puzzle'.

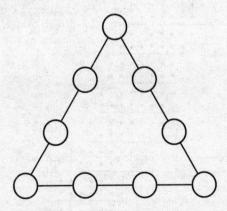

315. How can the digits 1 to 9 be placed in these circles, so that each side of the triangle sums to 20? How can the sum be made to equal 17?

316. The Axle Poser Why does the front axle of a cart usually wear out faster than the back?

317. Cutting the Cube It is easy to cut a cube into twenty-seven smaller cubes by slicing it twice vertically, twice horizontally and twice from back to front. This makes a total of six slices.

Suppose, however, that having made one or more slices you are allowed to rearrange the pieces as you choose, before making the next slice. Is it now possible to dissect the cube into twenty-seven smaller cubes in only five slices?

318. Counting To and Fro Mike was standing in the doorway of his house, counting the people passing in both directions. Tom was walking up and down the road, counting all those he passed, in either direction. After an hour, they met and compared the number they had counted. Who had counted most?

319. A Terminating Division 'In the terminating division shown below, the dots represent unknown digits, and in the entire division the location of only seven digits 7 are known. However, it is also possible that a dot represents a digit 7.'

Reconstruct the completed sum.

320. The Overhanging Bricks (1)   A large number of identical rectangular bricks are piled in a vertical column, so that each brick is immediately over the brick below. What is the greatest overhang that can be achieved by sliding the bricks over each other, parallel to their longest sides?

321. The Overhanging Bricks (2)   With the same supply of bricks and a table-top, what is the greatest projection possible, over the edge of the table, using only four bricks?

The bricks may be arranged in any manner, but the maximum projection is measured as the maximum distance from the edge of the table to the end of a brick.

322. A Leap in Age   The day before yesterday I was 13 years old. Next year I shall be old enough to get married. When is my birthday, and what is the date today?

323. The Cylindrical Hole   A hole 6 inches long is drilled through the centre of a solid sphere. What is the volume of the sphere remaining?

324. The Lost Pound   Three diners on finishing their meal are presented with a bill for £30, which they agree to split between them. They each gave the waiter £10, not knowing that the waiter had rechecked the bill and found that it was only £25. At this point it occurred to the waiter that £5 would not divide equally between the three, and anyway they did not know that there had been a mistake. So he returned to the table, apologized and gave each diner £1, keeping the other £2 for himself.

The diners have each now paid £9, making £27 in all, and the waiter has £2 in his pocket, a total of £29. Yet they originally gave the waiter £30. Where has the missing pound gone?

325. The Submerged Balance   A lump of lead is being weighed on a balance, placed on the flat bottom of a basin, against several iron weights. When the stone and weights balance exactly, the balance is submerged by filling the basin with water.

Will it stay balanced? If not, which way will it tip?

326.   'This diagram [overleaf] shows a cube with a piece cut off. Your problem is this: can you tell from this diagram if the slice ABCD could be a flat slice? That is, could the points, A, B, C and D lie in a plane?'

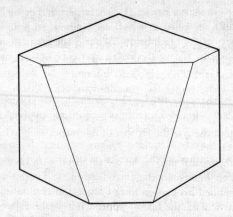

327. The neighbouring countries of Moria and Monia were both jealous of each other's economic success, so when the government of Moria announced that in future a Monia dollar would be worth only 90 Moria cents, the government of Monia retaliated by announcing that in future each Moria dollar would, likewise, be worth only 90 Monia cents.

At once, a bright young spark who lived in Monia, near the border, bought a 10 cent doughnut with his Monia dollar, exchanged his 90 cents for a Moria dollar, crossed into Moria and bought himself a soft drink for 10 cents. He then exchanged his 90 Moria cents for a Monia dollar, leaving him with the money he started with.

Who paid for the doughnut and the soft drinks?

328. The Matching Birthdays   How many people must be gathered together in the same room, before you can be certain that there is a greater than 50/50 chance that at least two of them have the same birthday?

The surprising result was first noted by the mathematician Harold Davenport.

329. Mastering the Masters   Mary was a delightful child with a precocious interest in chess, so she was delighted when her father arranged for two chess masters to visit her house. She was even more pleased when they both agreed to play a game with her, and she was ecstatic when she won one of the games, as conclusively as she lost the other. Given that Mary was actually too weak to beat either of the masters in a million years, what happened?

Hubert Phillips

Hubert Phillips was a prolific composer of all kinds of puzzles who had some strikingly original ideas, as well as producing many variations in particular themes, especially inference and deduction.

He often wrote under the pseudonyms of 'Caliban' in the *New Statesman*, and 'Dogberry' in the *News Chronicle*. He was also editor of *British Bridge World* and twice captain of England at contract bridge, a humourist and the creator of the Inspector Playfair detection mysteries.

In his mathematical puzzles he collaborated with his friend Sydney Shovelton, and others. In the Introduction to *The Sphinx Problem Book* he explained his conception of a good inferential-mathematical puzzle: in particular, 'the line of approach must be as well concealed as possible. I have put a good deal of thought into the construction of problems which at first blush appear to be insoluble, through inadequacy of the data. The invention of such exercises, and the solving of them, both give a great deal of pleasure, since their construction can involve – and in my view should have reference to – principles of artistry which embody an aesthetic of their own.'

In *Question Time* he also commented on his criteria for a good puzzle: 'Does its statement involve, not only the labour of working out the answer (which for many has a very slight appeal) but also the excitement of first discovering how the answer is to be arrived at? My main pleasure, in constructing puzzles, lies in seeking to provide this "kick".'

I am reminded of the great mathematician G. H. Hardy's talking in his *Mathematician's Apology* of 'the puzzle columns in the popular newspapers. Nearly all their immense popularity is a tribute to the drawing power of rudimentary mathematics, and the better makers of puzzles, such as Dudeney or "Caliban", use very little else . . . what the public wants is a little intellectual "kick", and nothing else has quite the kick of mathematics.'

The first two problems are based on an idea which first occurs in his 'Problems for Young Mathematicians' in *The Playtime Omnibus*:

330. What Colour was the Bear?   A man out hunting, spotted a bear due east. Taken by surprise, he ran directly north, and turned to see that the bear had not moved. Steadying himself, he took aim and shot it, by aiming due south. What colour was the bear?

331. A Roundabout Journey Mrs Agabegyun left her house one morning, and walked 5 miles south. She then turned east and walked another 5 miles. Finally she turned again and walked 5 miles due north, arriving back at her house.

Where does she live?

Phillips also originated the 'liars and truth-tellers' theme, which has subsequently lent itself to endless variation:

332. Red and Blue 'The island of Ko is inhabited by three different races – the Blacks, who invariably tell the truth; the Whites, who invariably lie; and the Muddleds, who tell the truth and lie alternately (though one cannot tell, in talking to a Muddled, whether his first remark is truthful or the reverse).

'In a certain school in Ko, a Black, a White and a Muddled were sitting side by side – in what order is not known. An inspector came in carrying a number of cards, some of them red and some blue. Taking a card at random, he asked each of the youngsters in turn: "What colour is this?" Then, taking a second card, he put to each of them in the same order, the same question. The six answers he received were:

(1) Blue, (2) Blue, (3) Red, (4) Red, (5) Blue, (6) Blue.'

What colour were the cards chosen by the inspector?

The next puzzle has also led to many variants, one of which follows immediately:

333. 'Two schoolboys were playing on the toolshed roof. Something gave way, and they were precipitated, through the roof, on to the floor below.

'When they picked themselves up, the face of one was covered with grime. The other's face was quite clean. Yet it was the boy with the clean face who at once went off and washed.

'*How is this to be explained?*'

334. The Three Wise Men Three Wise Men were taking a nap when a practical joker marked a cross on the forehead of each, with charcoal. The joker then hid behind a pillar and yelled loudly. At once they awoke and each started laughing at the plight of the others, until suddenly one of them stopped laughing and felt his own forehead, having realized that he was a victim of the same trick.

How did he draw this conclusion?

335. The Ship's Ladder 'The good ship *Potiphar* lay at anchor in Portsmouth Harbour. An interested spectator observed that a ladder was dangling from her deck; that the bottom four rungs of the ladder were submerged; that each rung was two inches wide and that the rungs were eleven inches apart. The tide was rising at the rate of eighteen inches per hour.

'*At the end of two hours, how many rungs would be submerged?*'

The next problem is a slight variant only of problem 521 in Dudeney's *536 Puzzles and Curious Problems*. It originally appeared in a Civil Service examination and created so much interest that the *New Statesman* persuaded Phillips to set similar problems in place of their crossword and bridge columns. Note the class distinctions involved in addressing the workers by their surnames alone, and the passengers with the prefix 'Mister'.

336. 'The driver, fireman and guard of a certain train were Brown, Robinson and Jones, and the passengers included Mr Brown, Mr Robinson and Mr Jones. Mr Robinson lived in Leeds; the guard lived midway between Leeds and London. Mr Jones' income is £400 2s 1d, and the guard's income is exactly one-third of his nearest passenger neighbour. The guard's namesake lives in London. Brown beats the fireman at billiards. What is the name of the engine-driver?'

337. Stritebatt '"Had a good season?" I enquired of my cricketing friend, Stritebatt.

'"Fairish. I finished with an average of exactly 30."

'"For how many innings?"

'"I forget. But we seldom had more than one in an afternoon, you know. And I've only been able to play on Saturdays."

'"Thirty is pretty good," said I.

'"Not bad," said Stritebatt. "But I was Not Out several times, you know. My friend Smith worked out that if I'd scored another dozen, in each of my Not Out innings, my average would have been 35."

'"Any good scores?" asked I.

'"Nothing special. Two blobs. Otherwise, my lowest score was 17. By the way, I had no two scores the same, apart from the blobs, I mean. And all my best scores were Not Out ones."

'*What was Stritebatt's highest score?*'

338. The Lodger's Bacon 'At Aspidistra House it is the landlady's custom to put the breakfast bacon on a dish before the fire, so the lodgers may help themselves as they come down in the morning. On this particular morning, all had shared alike, there would have been a whole number of rashers each, but Smith, who came down last, found only half a rasher left for him. Jones is very regular, and always takes one rasher; Robinson takes his fair share of what remains when he comes down; Brown is greedy, and takes his fair share and then half a rasher extra; Evans likes three rashers, but, being superstitious, always leaves one at least for those who follow him.

'How many rashers did Evans take?'

339. Falsehoods   'Messrs Draper, Grocer, Baker and Hatter are (appropriately enough) a draper, grocer, baker and hatter. But none of them is the namesake of his own vocation.

'When I tried to find out who is who, four statements were made to me: (1) "Mr Draper is the hatter." (2) "Mr Grocer is the draper." (3) "Mr Baker is not the hatter." (4) "Mr Hatter is not the baker." But clearly there was something wrong here, since Mr Baker is not the baker.

'I subsequently discovered that three of the four statements made to me are untrue.

'Who is the grocer?'

340. 'Alice was Disconcerted'   '"What's 19 times 19?" asked the Red Queen.

'"361," said Alice.

'"Wrong," said the Red Queen. "The answer's 519."

'Alice was disconcerted. They had forgotten to tell her that, just to make things difficult, each digit in this multiplication represented a different one.

'What number does "19" represent?'

341. Moulting Feathers   'Last summer I spent a week or so in the little-known village of Moulting Feathers, in Dumpshire. Its social centre is the local Bird Fanciers' Club.

'The club has seven members. Each is the owner of one bird. And each owner is, strange to say, the namesake of the bird owned by one of the others.

'Three of the fanciers have birds which are darker than their owners' feathered namesakes.

'I stayed with the human namesake of Mr Crow's bird, from whose wife I collected most of the village gossip. Incidentally, only two of the fanciers – Mr Dove and Mr Canary – are bachelors.

'Mr Gull's wife's sister's husband is the owner of the raven – the most popular of the seven birds. The crow, on the other hand, is much disliked; "I can't abide him," said his owner's fiancée.

'Mr Raven's bird's human namesake is the owner of the canary, while the parrot's owner's feathered namesake is owned by the human namesake of Mr Crow's bird.

'*Who owns the starling?*'

342. Windows 'Sir Draftover Grabbe, Chancellor of the Exchequer, conceived the unoriginal plan of imposing a tax on windows. A window having 12 square feet of glass paid a tax of £2 3s; a 24-square-foot window paid £3 1s; a 48-square-foot window paid £4 17s.

'*What do you suppose was the basis of the tax?*'

Looking-glass Zoo   This puzzle was composed by the famous astrophysicist and philosopher Sir Arthur Eddington, and originally published by Hubert Phillips in *Question Time*:

343. 'I took some nephews and nieces to the Zoo, and we halted at a cage marked

> Tovus Slithius, male and female.
> Borogovus Mimsius, male and female.
> Rathus Momus, male and female.
> Jabberwockius Vulgaris, male and female.

The eight animals were asleep in a row, and the children began to guess which was which. "That one at the end is Mr Tove." "No, no! It's Mrs Jabberwock," and so on. I suggested that they should each write down the names in order from left to right, and offered a prize to the one who got most names right.

'As the four species were easily distinguishable, no mistake would arise in pairing the animals; naturally a child who identified one animal as Mr Tove identified the other animal of the same species as Mrs Tove.

'The keeper, who consented to judge the lists, scrutinized them carefully. "Here's a queer thing. I take two of the lists, say, John's and Mary's. The animal which John supposes to be the animal which Mary supposes to be Mr Tove is the animal which Mary supposes to be the animal which John supposes to be Mrs Tove. It is just the same for every pair of lists, and for all four species.

'"Curiouser and curiouser! Each boy supposes Mr Tove to be the animal which he supposes to be Mr Tove; but each girl supposes Mr Tove to be the animal which she supposes to be Mrs Tove. And similarly for the other animals. I mean, for instance, that the animal Mary calls Mr Tove is really Mrs Rathe, but the animal she calls Mrs Rathe is really Mrs Tove."

'"It seems a little involved," I said, "but I suppose it is a remarkable coincidence."

'"Very remarkable," replied Mr Dodgson (whom I had supposed to be the keeper) "and it could not have happened if you had brought any more children."

How many nephews and nieces were there? Was the winner a boy or a girl? And how many names did the winner get right?

344. Wheels around Wheels   How many times does a coin rotate in rolling completely about another coin, of the same size, without slipping?

345. Covering a Chessboard   If two squares are removed from a chessboard, one from each end of one of the long diagonals, can the squares that remain be covered by thirty-one dominoes, each large enough to exactly cover a pair of adjacent squares?

346. Too Many Girls   In a far off land where warfare had raged for many years, the number of men was too few for the number of women who wished to marry them. While nothing could be done immediately about this sorry state of affairs, the King was determined that in future there should be more boys born and fewer girls, in anticipation of the ravages of war.

With this aim in mind, he decreed that every woman should cease to bear children as soon as she gave birth to her first daughter, reasoning that while there would be some families which would have only one daughter, or even one son and one daughter, there would be many with several sons followed by a single daughter, producing an overall surplus of sons.

Where did his ingenious scheme go wrong?

347. Forty Unfaithful Wives   'The great Sultan was very much worried about the large number of unfaithful wives among the population of his capital city. There were forty women who were openly deceiving their husbands, but, as often happens, although all these cases were a matter of common knowledge, the husbands in question were ignorant of their wives' behaviour. In order to punish the wretched women, the sultan issued a proclamation which permitted the husbands of unfaithful wives to kill them, provided, however, that they were quite sure of the infidelity. The proclamation did not mention either the number or the names of the wives known to be unfaithful; it merely stated that such cases were known in the city and suggested that the husbands do something about it. However, to the great surprise of the entire legislative body and the city police, no wife killings were reported on the day of the proclamation, or on the days that followed. In fact, an entire month passed without any result, and it seemed the deceived husbands just did not care to save their honour.

'"O Great Sultan," said the vizier, "shouldn't we announce the names of the forty unfaithful wives, if the husbands are too lazy to pursue the cases themselves?"

'"No," said the Sultan. "Let us wait. My people may be lazy, but they are certainly very intelligent and wise. I am sure action will be taken very soon."

'And, indeed, on the fortieth day after the proclamation, action suddenly broke out. That single night forty women were killed, and a quick check revealed that they were the forty who were known to have been deceiving their husbands.

'"I do not understand it," exclaimed the vizier. "Why did these forty wronged husbands wait such a long time to take action, and why did they all finally take it on the same day?"'

•

348. What Moves Backwards
When the express from Bristol to London is thundering towards London, some parts of the train are, at one moment, moving towards Bristol. Which parts?

349. The Backwards Bicycle
A bicycle is supported vertically, but is free to move forwards and backwards when the handlebars are pushed. One pedal is at its lowest point and the other is at its highest point.

If a string is attached to the lower pedal, and pulled backwards, will the bicycle move forwards or backwards?

350. The Heads of Hair
There are at least 50 million people living in the United Kingdom, and no human being has more than a million hairs on their head. What is the least number of inhabitants of the United Kingdom who must have, according to the information, exactly the same number of hairs on their head?

351. Quickies   (a) 'Have you ever seen anyone running along the pavement and placing their feet on the ground in this order: right foot, right foot, left foot, left foot, right, right, left, left . . . ?'

(b) 'By suitably placing a six-inch square over a triangle I can cover up to three-quarters of the triangle. By suitably placing the triangle over the square, I can cover up to one-half of the square. What is the area of the triangle?'

(c) 'When is it polite to overtake, or pass, on the inside only?'

(d) '"Don't forget you owe me five pence!" said Fred.
'"What!" replied Tom, "Five pence isn't worth bothering about."
'"All right then," said Fred, "you can give me ten pence."
'What is the logic behind Fred's reply?'

352. An Amazing Escape   'Archaeologists, more than most scientists, destroy cherished myths with every discovery they make. When they claim, however, that the Labyrinth which trapped Theseus was merely the rooms of a palace with which he was unfamiliar, they are going

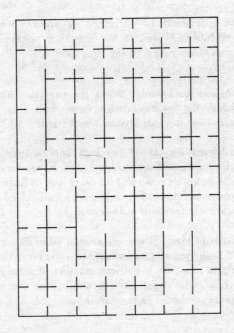

too far. They fail to appreciate that when you are being hotly pursued by a Minotaur, you *must* take every opportunity to turn left or right to escape the beast, however quickly you might otherwise escape in a straight line. Naturally Theseus, who entered by the south entrance to the palace, wanted to get to his beloved Ariadne, who was waiting just outside the north entrance, as quickly as possible. What was his shortest route if he was to evade the Minotaur?'

353. Speedy Gonzales 'The other day, travelling by London underground, I dashed on to the platform just as my train was moving out. I caught the next one, and left the exit of my destination station at exactly the same time as I would have done had I not missed the first train. Both trains travelled at the same speed, no acts of God were involved, and I didn't have to rush to make up for lost time. Explanation please?'

354. An Intimate Affair 'At an intimate little soirée given by Lady What's-her-name the other evening, each man danced with exactly three women and each woman with exactly three men. What is more, each pair of men had exactly two dancing partners in common. An admirable arrangement which pleased Lady What's-her-name no end and also gives the reader enough information to discover exactly how intimate that soirée was. How many people attended?'

355. Rice Division 'Mr and Mrs Lo Hun were poor peasant farmers, so when Mrs Lo Hun accidentally smashed the measuring bowl which she used for measuring out the rice, she was very upset. Fortunately, her husband was skilled in the traditional art of sword-fighting, and she brightened considerably when he took a strong cardboard box of rectangular shape and, with the minimum necessary number of clean plane sword cuts, produced a substitute for her bowl which actually measured out one, two, three or four measures of rice, according to her choice.

'How many cuts did her husband make, and what shape was the final article?'

356. A Mon-ster Puzzle The illustration overleaf shows a *mon*, a Japanese family crest. Janet wanted to show it in her project on Japanese history, and she was just about to cut two equal squares of gummed paper, one black and one white, into quarters and stick them down so that they overlapped in sequence, when it struck her that eight pieces might not be necessary. Indeed, they were not, and

eventually she managed to make the *mon* by using fewer pieces. How many separate pieces did she use?

357. Knot these Cubes   What is the shortest knot that can be tied in three dimensions using only face connected cubes? All the cubes are the same size, the knot must be continuous with no loose ends, and the cubes must be connected by complete faces.

358.   These twelve matches form one square and four triangles. How can half of them be moved to form one triangle and three squares?

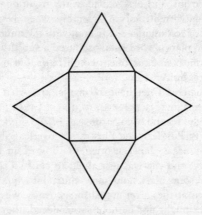

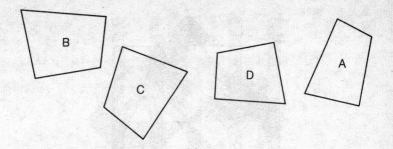

359. Two Squares in One Each of these four pieces has two right-angles, so it is hardly surprising that they can be fitted together to form a square in more than one way. In how many, precisely?

360. Half-hearted Betting Major Watson, despite being short of the readies, was feeling in a good mood, so he decided to gamble with his friend Butterworth on the toss of a coin. Starting with just £1, he bet six times, each time wagering half the money he had at the time.

If he won three times and lost three times, how much did he win or lose in the end?

361. Chopsticks 'A firewood merchant had a number of blocks to chop up for firewood. He chopped each block into eleven sticks. Assuming that he chopped at the average rate of forty-five strokes per minute, how many blocks would he chop up in twenty-two minutes?'

362. The Convivial Visitor '"There are only four pubs in this village," the visitor was informed, "one in each street. The village's four streets meet at the crossroads at right-angles. This street is the High Street.

'"To reach the Blue Boar from the Griffin you must turn left. To reach the Dragon from the Red Lion you have to turn right."

'The visitor entered three of the pubs; he arrived at the crossroads three times during this pilgrimage, turning left the first time, going straight across the second, and turning right the third time. He spent the night at the Blue Boar.

'Which pub stands in the High Street?'

363. Siding by Siding These five locomotives have to be driven into their respective sheds, marked with their number. Moving an engine into a shed and out again, without moving any other engine in the

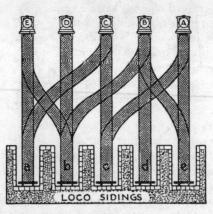

meantime, counts as one move, so for example moving A into e, out to B and into d, would count as just one move.

The puzzle is to do this as efficiently as possible.

364. Triangles within Triangles The sides of a triangle have each been divided into quarters, and each vertex joined to one of the points of division, as shown in the figure.

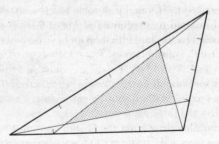

What is the area of the triangle in the centre?

365. A Unique Number What is the unique whole number whose square and cube between them use up each of the digits 0 to 9, once each?

366. The Picnic Ham 'Three neighbours gave $4 each and bought a ham (without skin, fat, and bones). One of them divided it into three parts asserting that the weights were equal. The second neighbour declared that she trusted only the balance of the shop at the corner. There, it appeared that the parts, supposed to be equal, corresponded to the monetary values of $3, $4 and $5, respectively. The third partner decided to weigh the ham on her home balance, which gave a still different result. This led to a quarrel, because the first woman kept insisting on the equality of her division, the second one recognized only the balance of the shop, and the third only her own balance. In what way is it possible to settle this dispute and to divide these pieces (without cutting them anew) in such a way that each woman would have to admit that she had got at least $4 worth of ham if computed according to the balance which she trusted?'

367. Giants and Midgets
Unlike the Household Guards, Major Mason's Mercenaries were a right shower, ranging from tall and skinny, to short and fubsy, not forgetting the enormous Corporal Gut and the minuscule Private Git.

One afternoon, Major Mason decided to divide them into three separate groups, the large, the middling and the small, who could parade separately without provoking the jeers of the locals.

He first assembled them in a rectangular array, and instructed the tallest man in each row to step out. From these he chose the shortest of the tall and announced that he and all those taller than him, would form the first new platoon.

The men returned to their initial positions, and the smallest man in each column was ordered to step out, and Major Mason picked out the tallest of the shortest, declaring that he and all the soldiers shorter than he would form the third platoon. The remainder would form the second platoon.

To the Major's rage and disgust, not only were there apparently no men at all in the second platoon, but Private Ponce claimed that he was in both the first and third platoons.

The Major naturally flogged every man jack of them severely, and declared that they would do it again, *and properly this time*. What was the result?

368. Near Neighbours
Trevor the travel agent has a map of Europe on which every major town is joined to the town nearest to it. The distances between towns are always different, when measured sufficiently accurately.

What is the largest number of other towns to which any one town can be connected?

369. Guarding the Gallery The new art gallery has twenty walls, each wall being at right-angles to its adjoining walls. Without knowing the precise design of the gallery, what is the least number of guards that will guarantee that all the walls can be kept under observation all the time?

370. Batty Batting Frank had an excellent first half of the season, averaging comfortably more runs per innings than Paul. Moreover he had started the second half of the season very well, and he looked forward to once again picking up the club trophy for best overall batting average.

At the end of the second half of the season, Frank had indeed once again beaten Paul's average, yet for the whole season, to Frank's disgust, Paul was ahead, and took the trophy. How was this possible?

371. The Bouncing Billiard Ball A mathematical billiard table is in the form of a rectangle with integral sides, and just four pockets, one in each corner. A ball shoots out of one pocket at angles of 45° to the sides. Will it bounce round the table for ever, or end up in one of the other pockets?

372. Back to the Start A billiard ball is struck without side so that it strikes all four cushions and returns to its starting position.

In what direction is it struck, and how far does it travel?

373. Lies, Almost All Lies Here are ten numbered statements. How many of them are true?

    1   Exactly one of these statements is false.
    2   Exactly two of these statements are false.
    3   Exactly three of these statements are false.
    4   Exactly four of these statements are false.
    5   Exactly five of these statements are false.
    6   Exactly six of these statements are false.
    7   Exactly seven of these statements are false.
    8   Exactly eight of these statements are false.
    9   Exactly nine of these statements are false.
   10   Exactly ten of these statements are false.

374. Bookworm A bookworm, feeling very hungry, is delighted to come across the three volumes of Dr Johnson's great *Dictionary of the English Language* standing on a shelf. Starting from the front cover of the first volume, it bores its way through to the back of the back cover of the third volume.

If the front and back covers of each volume are $\frac{1}{2}$ cm thick and the pages of each volume are 7 cm thick, how far does the bookworm bore?

375. OH-HO How many moves are required to transform this H into the O, if a move consists of sliding one coin to touch two others, without moving any of the other coins?

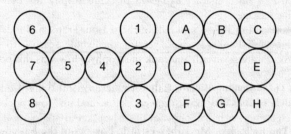

When you have changed the H to O, how many moves does it take to get back from O to H?

376. Inverted Triangle This triangle contains ten coins. What is the smallest number that must be moved to make the triangle point downwards?

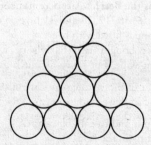

377. Paradoxical Dice Alan, Barry and Chris were playing at dice, using three fair dice which they had each marked with their own special numbers. Alan consistently beat Barry, and Barry's dice consistently beat Chris's. What was surprising was that Chris's dice nevertheless consistently beat Alan's.

How was this possible?

378. The Obedient Ray Two mirrors are joined at a fixed angle at O, and a ray of light is shone into the angle between them, parallel to one of the mirrors. It bounces a number of times, strikes the lower mirror at right-angles at X, and then re-emerges along its original path.

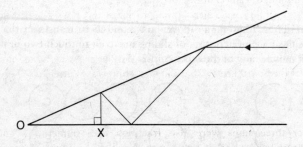

What is the distance between the original ray and the lower mirror, to which it is parallel?

379. The Balloon 'Mr Tabako's little boy sits in the back seat of a closed motor-car, holding a balloon on a string. All the windows of the car are closed tight. The balloon is full of coal gas and is tethered by a string, which prevents it from touching the roof of the car.

'The car turns left at a roundabout. Does the balloon swing left, swing right, stay upright, or do something else? And why?'

380. Which Contains the Beer? A grocer has six barrels of different sizes, containing 15, 16, 18, 19, 20 and 31 litres. Five barrels are filled with wine and only one is filled with beer.

The first customer bought two barrels of wine, and a second customer also bought wine, but twice as much as the first. Which is the beer barrel?

381. Blackbirds
> 'Twice four and twenty blackbirds
> Were sitting in the rain.
> Jill shot and killed a seventh part.
> How many did remain?'

382. The Two Girlfriends John is equally devoted to his two girl-friends, one of whom lives uptown and the other downtown. He therefore decides to catch the first bus to arrive, whichever direction it is going in. Since all the buses run at equal intervals, and his own times of arrival at the road are quite random, he looks forward to visiting each girl with equal frequency, yet he soon finds out that he is seeing one far more often than the other. Why?

383. Confounded Cancellation Mr Peebles believed in giving his pupils responsibility, so when he had to leave the class one day he instructed Jones Minor to go to the board and write up some simple fraction sums which the rest of the class were to do until he returned.

On his return, he found the class rolling in the aisles with laughter. Looking at the board he saw that the first 'sum' was written as

$$\frac{1\!\!\!/6}{6\!\!\!/4} = \frac{1}{4}$$

The other three sums were also fractions with numerators and denominators below 100, and each was simplified by Jones Minor in the same absurd – but in this case 'correct' – manner. What were they?

384. Back to Back These cards are used in a simple psychological test. Every card certainly has a letter on one side and a number on the other. I make the additional claim to you that if you look at the letter on one side of a card and see that it is a vowel, then you can be certain that the number on the other side is even.

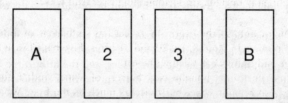

How many cards must you turn over to check whether my additional claim is correct?

385. The Cigarette Ends A tramp collecting cigarette ends from the street can make a new cigarette out of four ends. He collects in one morning, thirty-two ends. How many cigarettes can he smoke that day?

386. Coin Catch I have only two coins in my pocket. They add up to 15 pence and yet one of them is not a 10 pence piece. What denominations are they?

387. Beer from a Can   You are drinking beer from a can. When the can is full, the centre of gravity of the beer and can together will be in the centre of the can, as near as makes no difference. As you start to drink, the centre of gravity falls, but by the time the can is empty it is back to the centre of the now-empty can.

At what point did the centre of gravity reach its minimum position?

388. Long-playing Poser   How many grooves are there on a standard long-playing record?

389. The Lily in the Pond   A water lily doubles in size, that is, in the area of the leaf lying on the surface of the pond, every 24 hours. If it takes 30 days to cover the pond completely, after how many days did it cover exactly one half of the pond?

390. A Shaking Result   At a recent conference, most of the delegates, but not all, shook hands with most of the other delegates on arrival, and again on leaving. Why was the number of delegates who shook hands an odd number of times necessarily even?

391. The Hotel Reception   A group of travellers, seven in number, arrived at a hotel and asked to be put up for the night. The manager actually only had six rooms available, but he promised to do what he could.

First he put the first man in the first room and asked another man to wait there for a few minutes. He then placed the third man in the second room, the fourth in the third room, and the fifth in the fourth room. Finally he placed the sixth man in the fifth room and went back for the seventh man, who he placed in the sixth room. OK?

392. An Irrational Number   'Show, by a simple example, that an irrational number raised to an irrational power need not be irrational.'

393. Horseshoe Dissection   How can a horseshoe be cut into six separate pieces with just two cuts?

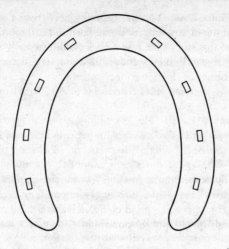

394. A Price Poser

'How much will one cost?'

'Thirty pence.'

'How much will fifteen cost?'

'Sixty pence.'

'Thank you, I'll take one hundred and sixteen.'

'That will be ninety pence, Madam.'

Explain, please.

395. Hula Hoop 'Consider a vertical girl whose waist is circular, not smooth, and temporarily at rest. Around her waist rotates a hula hoop of twice its diameter. Show that after one revolution of the hoop, the point originally in contact with the girl has travelled a distance equal to the perimeter of a square circumscribing the girl's waist.'

396. Up and Down John Smith leaves home every morning, from his flat at the top of a tower block, and takes the lift to the ground floor, walks to the bus stop and catches the bus.

On the way home, however, he gets off the bus, walks to the tower block entrance, takes the lift to the seventh floor and then walks the rest of the way. Why? It may help you to know that he is extremely healthy, and is not in need of exercise.

397. Three into Two You have a frying pan which will take only two slices of bread at a time, and you wish to fry three slices, each on both sides. Since each slice takes 20 seconds for each side, you can certainly fry them all in 80 seconds, by doing two pieces together and then the third.

But can you fry them more efficiently?

398. The Jigsaw Puzzle 'In assembling a jigsaw puzzle, let us call the fitting together of two pieces a "move", independently of whether the pieces consist of single pieces or of blocks of pieces already assembled. What procedures will minimize the number of moves required to solve an *n*-piece puzzle? What is the minimum number of moves needed?'

399. The Ladder and the Box A ladder, 4 metres long, is leaning against a wall in such a way that it just touches a box, 1 metre by 1 metre, as in the figure. How high is the top of the ladder above the floor?

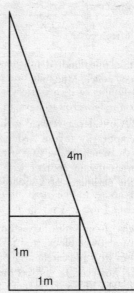

400. The Crossed Ladders   Two ladders, 20 and 30 feet long, lean across a passageway. They cross at a point 8 feet above the floor. How wide is the passage?

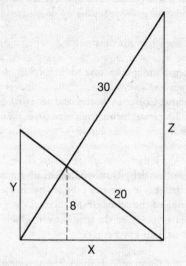

401. Odd Corners   The regular tetrahedron and the regular dodecahedron both have vertices at which an odd number of edges meet, but they have an *even* number of such vertices.

Is it possible for a polyhedron to have an odd number of vertices at which an odd number of edges meet?

402. Where does Mr Jones Live?   Mr Jones has moved to a new house in a rather long street, and has noticed that the sum of the numbers up to his own house, but excluding it, equals the sum of the numbers of his house to the end house in the road. If the houses are numbered consecutively, starting from 1, what number does Mr Jones live at?

403. The Knockout Tournament   If the number of players entered for a knockout tournament is a power of 2, for example 8, 16 or 32, then it is easy to arrange the pairings and it is obvious how many matches will take place in each round.

What happens if there is a different number of entrants? In particular, how many matches will have to be played if thirty-seven players enter a knockout?

404. The Breakfast Egg Mr Oval started every day with an egg, lightly boiled, with a slice of toast, yet he never bought an egg, neither borrowed nor stole his eggs and did not keep chickens. Please explain!

405. A Riddle Two legs sat on three legs when along came four legs and stole the one leg, whereupon two legs picked up three legs and threw it at four legs, and got his one back. Explain, please.

406. The Bottle and Cork A bottle and its cork cost 21 pence and the bottle costs 20 pence more than the cork. What is the cost of each?

407. The Mixed-up Labels You are given three boxes containing, respectively, chocolate drops, aniseed balls, and a mixture. Unfortunately, every jar has been wrongly labelled with the label that ought to have gone on one of the other jars.

What is the least you need do to discover which jar is which and restore the labels to their correct jars?

408. Knotted or not Knotted? Is this piece of rope genuinely knotted, or just in a tangle? In other words, what will happen if you pull the two ends apart? Will it tighten into a knot, or stretch into a straight line?

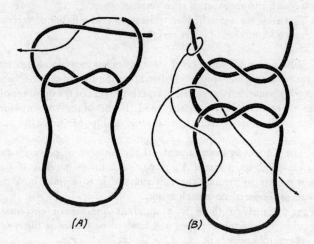

(A)   (B)

409. The Circle and Saucers 'Our table top is circular and its diameter is fifteen times the diameter of our saucers, which are also circular. How many saucers can be placed on the table top so that they overlap neither each other nor the edge of the table?'

410. Balls in a Box What is the size of the smallest cubical box which will just contain four balls, each ten inches in diameter?

411. The Problem of the Calissons *Calissons* are a French sweet, in the shape of two equilateral triangles edge to edge, which come packed in a hexagonal box. As you can see from the figure, the *calissons* can be pointing in any of three directions.

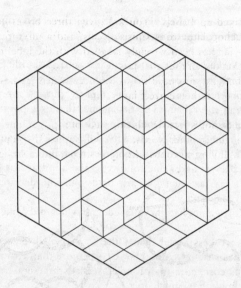

Your puzzle is to explain why the number of *calissons* pointing in each direction, in a fully packed box, must always be equal.

412. The Long Shot A hunter travelling by train to the forest carries with him his gun, which is 2.3 metres long. Unfortunately, the baggage regulations of the train company forbid any object more than 2 metres long. How does the hunter get round this rule?

413. Cigarette Extras A manufacturer produces boxes of cigarettes. Each box contains 160 cigarettes arranged in eight rows of twenty.

Assuming that the cigarettes completely fill the box, it is nevertheless possible to get more cigarettes into the box. How can this be
done, and how many extra can be accommodated?

414. A Moving Poser Place three coins in a row, like this, so that
each touches the next:

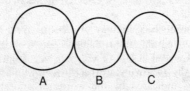

The puzzle is to move coin A so that it is between coins B and C,
without touching either B or C.

415. High Stakes 'Mike sat down and started shuffling the cards.
"What stakes?" he asked.

'"Let's make it a gamble," Steve replied, putting a few bills and
some coins in the table. "The first game, the loser pays 1 cent, the
second 2 cents, and so on. Double up each time."

'"Okay," laughed Mike, checking his cash, "I've got only $6.01
and I'm not playing more than ten games anyway."

'So they played, and game followed game until at last Mike stood
up. "That's my last cent I've just paid you," he declared, "but I'll
have my revenge next week."

'How many games had they played, and which did Mike win?'

416. Some are Less Equal than Others 'A pencil, eraser and notebook together costs $1. A notebook costs more than two pencils, and
three pencils cost more than four erasers. If three erasers cost more
than a notebook, how much does each cost?'

417. Pandigital Probability What is the probability that a ten-digit
number, that is, a number chosen at random between 1,000,000,000
and 9,999,999,999 inclusive, will have ten *different* digits?

418. The Broken Stick (1) A stick is broken into three pieces. What
is the probability that they will form a triangle?

To make what we have in mind a little clearer, let's say that two
points are chosen at random on the stick, each choice being independent of the other, and the stick is broken at those points.

419. The Broken Stick (2) A stick is broken into two pieces, at random. What is the average length of the shorter piece?

420 A Striking Clock When a grandfather clock strikes 6 o'clock, there are 15 seconds between the first and last strokes. How many seconds elapse between the first and last strokes when it strikes midnight?

421. Buried Treasure A treasure is buried somewhere along a straight road on which are four towns, the distances between them, in sequence, being 5, 8 and 11 miles.

A map gives the following instructions for finding the treasure. Unfortunately, as indicated by the words in quotation marks, the actual names of the towns have become illegible with age. Despite this difficulty, the site of the treasure can be located. How?

> Start at town 'squiggle' and go half of the way to 'splodge'.
> Then go one third of the way to 'can't read it' and finally travel
> one quarter of the way to 'illegible'.

422. Multiple Ages A man and his grandson have the same birthday. For six consecutive birthdays the man is an integral number of times as old as his grandson. How old is each at the sixth of these birthdays?

423. Grandfather and Grandson 'In 1932 I was as old as the last two digits of my birth year. When I mentioned this interesting coincidence to my grandfather, he surprised me by saying that the same applied to him too. I thought that impossible . . .'

What were their ages?

424. Diagonals of a Cube This figure shows two face diagonals of a cube. What is the angle between them?

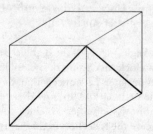

425. Equal Areas   How many regions of equal area can you see in this figure?

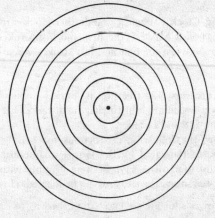

426. A Great Day for the Race   'Fred Bretts noticed that there were nine runners in the big race and asked his bookie what odds he was offering.

'"3–1 on Bonnie Lass, 4–1 on Golden Stirrup, 7–1 on Two's a Crowd, 9–1 on Greek Hero and 39–1 the field," he replied.

'Fred thought for a few moments and then astounded the bookie by placing a bet on each of the nine horses, all to win. No each-way nonsense for fearless Fred. And all on credit, of course.

'"You might as well give me my winnings now," said Fred.

'"The race hasn't been run yet, Sir," smiled the bookie.

'"That doesn't matter," said Fred. "When it has, you'll owe me £200."

'And he was right.

'How much did he stake on each horse?'

427. Find the Centre   How can the centre of a circle be found, accurately, by the use of a set-square only?

428. The Trisected Angle   'A confirmed angle-watcher one afternoon ruled two lines on a clock face to mark the angle then formed by the two hands. Some time later he noticed that the two hands exactly trisected the angle he had marked. In how short an interval could this have happened? And how soon after 3 o'clock could he have ruled his lines in order to observe the trisection in this short a time?'

429. Popsicle Polygons   The figure shows how five iced-lolly sticks, called popsicle sticks in the United States, can be used to make a triangle that can be picked up and handed round without falling apart.

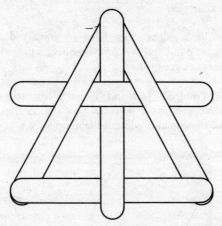

How many sticks are needed to make a regular hexagon?

430. Touching Three   What is the smallest number of pennies that must be placed on a table for each penny to touch exactly three others, if every coin is flat on the table?

431. Eight Heads and Eight Tails   Lay down sixteen coins, heads and tails alternately as shown. The problem is to rearrange the coins so that those in each vertical column are alike. Two coins only, may be touched.

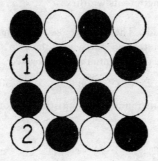

432. Falling on Edge 'How thick should a coin have to be to have a 1 in 3 chance of landing on edge?'

433. A Simple Angle How can an angle of 30° be constructed using only an unmarked ruler?

434. Bisecting the Segment You have an unmarked ruler, whose opposite edges are parallel. How can you bisect a given line segment, which is shorter than your ruler?

435. A Moving Problem Jack and Jill are moving to a new flat and their grand piano presents a potential problem. Fortunately, it will just pass round the corridor without being tipped on its end or being disassembled.

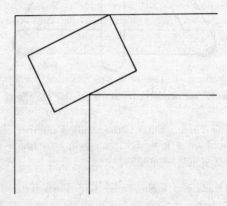

Given that its area, on looking down on it from above, is the largest possible which can be passed round the corner, what are the proportions of its length to its width?

436. Square and Add The number 3025 has the curious property that if you split it into two parts, add the two parts together and square the result, the original number is recovered:

$$30 + 25 = 55 \quad \text{and} \quad 55^2 = 3025.$$

What is the only other number, consisting of four different digits, with this property?

437. Striking a Balance 'Having a parcel to send by post, and being uncertain whether it was under 2 pounds (or 32 ounces) in weight, when it would go for sixpence, or heavier than that, which would mean ninepence, I borrowed some rather primitive scales from my landlady.

'The first weighing gave $28\frac{1}{4}$ ounces, well within the sixpenny range, but, being rather doubtful of the balance, which looked as if one arm was longer than the other, I tried weighing the parcel in the other scale-pan. This gave the weight as 36 ounces, thoroughly justifying my suspicions.

'How should I stamp my parcel?'

438. Dud Coins by the Boxful Mr Jones has plenty of coins, ten boxes of them in fact, but unfortunately one box contains duds which are all 2 gm short in weight. Even more unfortunately, he has forgotten which box contains the duds. If all the other boxes contain good coins, weighing 40 gm each, how many weighings on a weighing machine are necessary to decide which box has the duds?

439. The Problem of Twelve Coins 'Among twelve coins there is one at the most which has a false weight. With three weighings on an equal-arm balance, *but with no use of weights*, show how to establish whether there is a false coin, or not; and if so, which it is, and whether it is too light or too heavy.'

440. A Sound Bet? 'I will bet you one pound,' said Fred, 'that if you give me two pounds, I will give you three pounds in return.'

'Done,' replied Jack. Was he?

441. Sealed Bids 'Red and Black each stakes a 5 pence piece. Now each competes for this pool by writing down a sealed bid. When the bids are simultaneously revealed, the high bidder wins the stakes but pays the low bidder the amount of his low bid. If the bids are equal, Red and Black split the stakes.

'How much do you bid, Red?'

442. Sharing the Sandwiches Jones and Smith were sharing a journey, and when they felt hungry they prepared to share their sandwiches, of which Jones had brought five and Smith had brought three.

However, seeing a stranger, who turned out to be Mr Watson, eyeing their sandwiches, they offered to share them with him. Watson accepted, and they shared the sandwiches equally, after which Watson insisted on contributing £2 to the cost of his lunch.

Jones immediately suggested that they split the money in proportion to their contributions, Jones taking five parts, or £1.25, and Smith taking three parts, or 75p. But Smith objected, insisting that this would not be just. Who was right, and how much did each receive?

443. Pulling a Pint 'A stranger walked into a public bar, put ten-pence on the counter and asked for half a pint of beer. The barmaid asked whether he would like Flowers or I.P.A. The stranger asked for Flowers.

'Another complete stranger entered the bar, put tenpence on the counter and asked for half a pint of beer. Upon which the barmaid immediately pulled half of Flowers. How did she know what the second man, who was a stranger to her, wanted?'

444. A Square Chessboard How many squares are there on an 8 × 8 chessboard?

445. Mid-point with Compass Only You are given two points which may be thought of as the ends of a line segment, except that the line isn't there. How can you find the mid-point of the imaginary segment using only a pair of compasses? No ruler, no straight-edge are allowed, and no folding to get a straight line by stealth.

446. Tricky Tumblers Here are six tumblers, three full and three empty, arranged in a row.

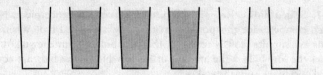

What is the smallest number of moves needed to leave the tumblers *alternately* full and empty? Every time a tumbler is picked up, that counts as a move.

447. Turning Tails There are eight ways to arrange three coins in a row, each coin showing either head or tail. Starting with three heads showing, and changing only one coin at a time, can you in just seven turns go through the entire sequence, ending up with three tails uppermost?

448. Little Pigley Farm, 1935 The first crossword appeared on 21 December 1913, in *The New York World*, but crosswords did not take off until 1924 when Simon and Schuster published a book of fifty puzzles. Crossword mania erupted, everyone jumped on to the bandwagon, and by the end of the first year 350,000 crossword books had been sold.

Naturally, cross-numbers were soon to follow. This puzzle is from *The Strand Problems Book* by W. T. Williams, who composed puzzles for *John O'London's Weekly*, and G. H. Savage, who published in *The Strand Magazine*.

Note: One of the 'across' numbers is the same as one of the 'downs'. This is the only case of identity, though one number in the puzzle (relating to something quite different) happens to be the area in roods of the rectangular field known as Dog's Mead. Equipped with this information and the homely items that follow, the reader is invited to discover that jealously guarded secret, the age of Mrs Grooby, Farmer Dunk's mother-in-law.

Readers may like to know that 1 acre was 4840 square yards, and 1 rood was one quarter of an acre. Also there were 20 shillings in £1 sterling.

Across
1. Area of Dog's Mead in square yards.

Down
1. Value in shillings per acre of Dog's Mead.

5. Age of Farmer Dunk's daughter, Martha.

6. Difference in yards of length and breadth of Dog's Mead.

7. Number of roods in Dog's Mead × 9 down.

8. Date (AD) when Little Pigley came into the occupation of the Dunk family.

10. Farmer Dunk's age.

11. The year when Mary was born.

14. Perimeter in yards of Dog's Mead.

15. The cube of Farmer's walking speed in miles per hour.

16. 15 ac. *minus* 9 down.

2. The square of Mrs Grooby's age.

3. Age of Mary, Farmer's youngest.

4. Value of Dog's Mead in pounds sterling.

6. Age of Farmer's firstborn, Ted, who will be twice as old as Mary next year.

7. Square of number of yards in breadth of Dog's Mead.

8. Number of minutes Farmer takes to walk 1⅓ times round Dog's Mead.

9. See 10 down.

10. 10 ac. × 9 down.

12. One more than sum of digits in column 2.

13. Length of tenure (in years) of Little Pigley by the Dunks.

449. Fours into Nine 'This is the grid for a children's crossword, in which no word of more than four letters will be used. Apart from this restriction, the grid will obey the usual rule that the black squares do not separate any part of the puzzle completely from the remainder.

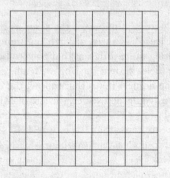

'What is the smallest number of black squares that must be filled in order to satisfy these conditions?'

450. A Common Libel

$$\frac{EVE}{DID} = .TALKTALKTALKTALK\ldots$$

This represents a common fraction written as a repeating decimal. What is the fraction?

451. 4 for Starters A number which ends in the digit 4, becomes 4 times larger when the 4 is removed from the end and placed at the front. What is the number?

452. Twelve Up You toss a dice with the usual numbers 1 to 6 on its faces, until the total exceeds 12. What is the most likely final total?

453. Mothers and Fathers First Only one large piece of cake remained, in the shape of a triangle. 'Equal shares for all!' announced Lilly, the tiniest.

'Agreed!' replied her mother. 'We shall all have pieces of exactly the same shape,' and so saying she cut the triangular cake into five pieces, all the same shape, two large and identical pieces for Father and herself, and three smaller identical pieces for the three children.

How much more cake did Father have than Lilly?

454. Arithmetic in Pictures Anyone can see that $5^2 + 10^2 = 11^2 + 2^2$, both being equal to 125. But can you demonstrate this by geometry? Specifically by dissecting each of these figures into the other, using of course as few pieces as possible?

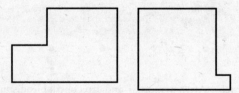

455. Dollars into Cents 'When Mr Smith cashed a cheque (for less than $100), the bank clerk accidentally mistook the number of dollars for the number of cents, and conversely. After Mr Smith had spent 68 cents, he discovered that he had twice as much money as the cheque had been written for. What was the amount for which the cheque had been written?'

456. Four through Nine   Taking your pencil, can you cross out all nine of these dots with four straight lines, without lifting your pencil off the paper?

457. Sixteen Out   Taking your pencil, cross out all sixteen of these dots, in a sequence of straight strokes, without lifting your pencil from the paper and ending up at the point where you started. How few strokes are required?

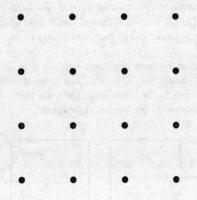

458. The Truel   'After a mutual and irreconcilable dispute among Red, Black and Gray, the three parties have agreed to a three-way duel. Each man is provided with a pistol and an unlimited supply of ammunition. Instead of simultaneous volleys, a firing order is to be established and followed until one survivor remains.

'Gray is a 100 per cent marksman, never having missed a bull's-eye in his shooting career. Black is successful two out of three times on the average, and you, Red, are only a 1/3 marksman. Recognizing the disparate degrees of marksmanship, the seconds have decided that you will be first and Black second in the firing order.

'Your pistol is loaded and cocked. At whom do you shoot?'

459. The Hurried Duellers 'Duels in the town of Discretion are rarely fatal. There, each contestant comes at a random moment between 5 a.m. and 6 a.m. on the appointed day and leaves exactly five minutes later, honour served, unless his opponent arrives within the time interval and then they fight. What fraction of duels lead to violence?'

460. Matching Matches Here is a row of fifteen matches. Arrange them in five groups of three each, by repeatedly moving one match so that it jumps over three matches.

461. Seven Up These seven cups have to be turned the right way up, but each move must consist of inverting three at a time. You can choose the three from anywhere in the line, they need not for example be adjacent, and a cup may be inverted on one move, inverted again on the next, and so on.

How many moves are necessary? How many moves would you need if the rules specified that four cups be inverted at each turn?

462. An Objectionable Rearrangement In Class 4C there are twenty-five desks arranged in five rows of five to form a square. On Tuesday they had a new teacher who instructed them to each move to a new desk, with the least fuss possible. That is, to move to the desk either directly in front or behind, or to the right or left of their old desk. Oh, I forgot to mention that Peaky Wilson was absent on Tuesday and his desk remained unoccupied.

On Wednesday Peaky had returned, and once again the teacher instructed all the students to each move to a new desk, under the same conditions as before. Peaky objected strongly, and ended up in the Headmaster's study, accused of insolence. Why?

463. Squaring the Cube   Your poser is to dissect a cube into a square using just four pieces. No, the solution is not to cut the cube into four identical square slices and use them as quarters of the square – because the 'square' will then be a shallow prism.

464. Cutting the Cake   Only Jane and her three closest friends are to cut her birthday cake. If they each make one vertical cut, what is the maximum number of pieces that they can cut?

If not all the slices have to be vertical, which is alright because the marzipan on top makes Patrick sick anyway, how many pieces of cake can the four of them cut?

465. A Hectic Week   'When the day after tomorrow is yesterday, today will be as far from Sunday as today was from Sunday when the day before yesterday was tomorrow. What day is it?'

466. Two proof readers are checking two copies of the same manuscript. The first finds thirty errors, and the second finds only twenty-four. When their completed proofs are compared, it turns out that only twenty errors have been spotted by both of them.

How many errors would you suspect remain, not detected by either of them?

467. How Many Mistakes?   How many mistakes are there in this sentence: 'This sentence contanes one misteak'?

What is the answer to the same question for this sentence: 'Their are three misteaks in this sentence'?

468. The Professor on the Escalator   'When Professor Stanislav Slapenarski, the Polish mathematician, walked very slowly down the down-moving escalator, he reached the bottom after taking fifty steps. As an experiment, he then ran up the same escalator, one step at a time, reaching the top after taking 125 steps.

'Assuming that the professor went up five times as fast as he went down (that is, took five steps to every one step before), and that he made each trip at a constant speed, how many steps would be visible if the escalator stopped running?'

469. Siting a Central Depot 'The street plan of a city consists only of straight streets intersecting at right-angles, and at an odd number of the junctions there are kiosks. The figure gives, as an example, a plan with ten streets and three kiosks. The occupants of the kiosks now wish to draw their wares from a common central depot. How should this be located so as to give a minimum total length for single trips to the depot from each individual kiosk? The breadths of the streets may be neglected.'

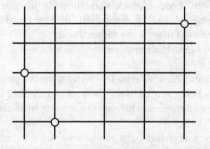

470. Nobel Prizes 'On the occasion of receiving his second Nobel prize, Dr Linus Pauling, the chemist, remarked that, while the chances of any person in the world receiving his first Nobel prize were one in several billion (the population of the world), the chances of receiving a second Nobel prize were one in several hundred (the total number of living people who had received the prize in the past) and that therefore it was less remarkable to receive one's second prize than one's first.'

What is the flaw in Professor Pauling's joke argument?

471. All Horses are the Same Colour Here is a proof that all horses are the same colour. One horse is certainly the same colour as itself. Now assume that the title statement is true of any set of N horses. Then it follows that it is true for any set of $N + 1$ horses, by the following reasoning:

Remove one horse from the set of $N + 1$ horses, to leave a set of N horses who are all, by our assumption, the same colour. Next, replace that horse and remove a different horse, to leave another set of N horses, all the same colour. By this argument, the two horses removed each have the same colour as the other $N - 1$ horses in the set. Therefore, all $N + 1$ horses have the same colour.

Where is the fallacy in this argument?

472. Father and Son   Mr Smith and his son were involved in a terrible accident at the factory where they worked. Mr Smith was killed outright, and his son was rushed to the emergency unit of the local hospital, and prepared for immediate surgery.

The surgeon on duty came into the operating theatre, saw the patient and exclaimed, 'That's my son, I can't operate!' and sent for a deputy.

Explain, please!

473. Father's Son   Lord Elphick was showing his guest the family portraits. Pointing to one, he remarked: 'Brothers and sisters have I none, but that man's father is my father's son.'

Who was represented in the portrait?

474. Three Teams, New Method   'What attracts people to watch football is goals being scored. And the authorities have been thinking for a long time of methods whereby the scoring of more goals might be encouraged.

'One suggestion that has been made involves a change in the way that points are awarded. The idea is that 10 points should be awarded for a win, 5 points for a draw, and 1 point for each goal scored, whatever the result of the match. Therefore even if you are losing 0–5 and have no hope of winning, a goal scored might make all the difference between promotion and non-promotion.

'This method was tried out on a small scale and its success can be judged from the fact that each side scored at least one goal in every match. There were only three sides playing and eventually they are all going to play each other once. A had scored 8 points, B had scored 14 points, and C had scored 9 points.

'*Find the score in each match.*'

475. Four Triangles   Four identical right-angled triangles have been added to a square, two pointing outwards, two inwards.

Why do the free vertices, marked with dots, lie on a straight line?

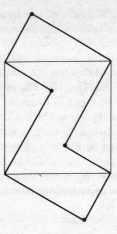

476. Three Digits What is the largest number that can be written with just three digits, using no other signs or symbols at all – and what are its last two digits?

477. The Five Couples 'My wife and I recently attended a party at which there were four other married couples. Various handshakes took place. No one shook hands with himself (or herself) or with his (or her) spouse, and no one shook hands with the same person more than once.

'After all the handshakes were over, I asked each person, including my wife, how many hands he (or she) had shaken. To my surprise each gave a different answer. How many hands did my wife shake?'

478. For the Love of a Good Woman Sir Pumphret and Sir Limpney both loved the Lady Isabel and resolved to have a race, the winner to take her hand in marriage. Knowing that Lady Isabel was opposed to all forms of competition, they chose to have a loser's race, the one whose horse came in last being the winner.

The first race was, predictably, a farce. They started very slowly, then went backwards, wandered off the course, and never came within sight of the finishing line.

The second race was quite different, both knights racing their mounts to the finishing line. Why?

479. The Egg Timer 'With a 7-minute hourglass and an 11-minute hourglass, what is the quickest way to time the boiling of an egg for 15 minutes?'

480. Kirkman's Schoolgirl Problem A schoolmistress is in the habit of taking her girls for a daily walk. The girls are fifteen in number, and on each walk are arranged in five rows of three, such that each girl might have two companions. The problem is to dispose them so that for seven consecutive days no girl will walk with any of her school-fellows more than once.

481. Common Tangents Here are three circles and six common tangents which meet in pairs at three points, X, Y and Z. It appears that XYZ is a straight line.

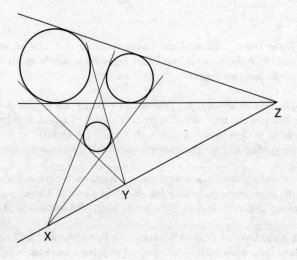

When this was first shown to Professor John Edson Sweet, a famous American engineer, Professor Sweet paused for a moment and said, 'Yes, that is perfectly self-evident.'

What was Professor Sweet's reasoning?

482. Polygon Products The figure shows how three regular dodecagons can be dissected, most elegantly, into pieces which assemble into one, larger, regular dodecagon.

Your problem is a little simpler. How can these three hexagon stars be cut up and reassembled into one star of the same shape?

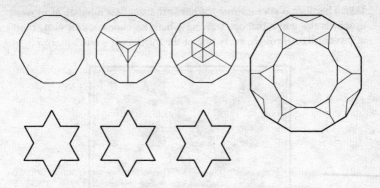

483. Dodecagon into Square This is a regular dodecagon and a square, of equal area.

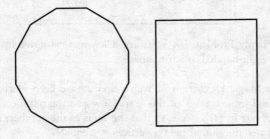

Your puzzle is to dissect each of them into six pieces that will reassemble to form the other. Because of a hidden – well, it's not that well hidden – feature of the two shapes, this is not as difficult as it might seem.

484. Knight's Tour On a small board, such as a 4 × 4 board, it is not possible to start at one square and visit every square, once and only once, making a knight's move (that is, the move of a knight in chess) each time. Indeed, on a 4 × 4 board, wherever you start, either four or six squares will be omitted from your tour.

What is the smallest rectangular board on which it is possible to do a complete tour?

Can you find a board on which it is possible to make a complete tour which has rotational symmetry, so that the tour remains unchanged when the board is given repeated quarter-turns?

485. The Bishop's Visitation   What is the smallest number of moves in which the white bishop, starting where he chooses, can visit, that is, pass into or through, every one of the thirty-two white squares?

486. A Handy Problem   If you turn a left-handed glove inside-out, will it be right-handed or left-handed?

487. The Magic Hexagram   Twelve circles have been placed at the vertices and intersections of this star. How can the numbers 1 to 12 be placed, one in each circle, so that the sums of the numbers in every row, and also the sum of the six vertices, are equal?

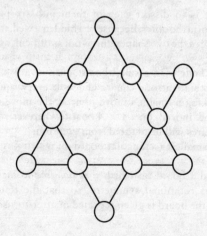

488. The Two Bookcases 'A room 9 by 12 feet contains two book-cases that hold a collection of rare erotica. Bookcase AB is 8½ feet long, and bookcase CD is 4½ feet long. The bookcases are positioned so that each is centred along its wall and one inch from the wall.

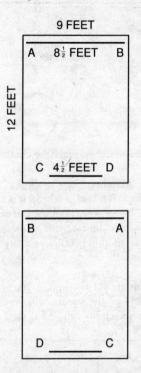

'The owner's young nephews are coming for a visit. He wishes to protect them and the books from each other by turning both book-cases around to face the wall. Each bookcase must end up in its starting position, but with its ends reversed. The bookcases are so heavy that the only way to move them is to keep one end on the floor as a pivot while the other end is swung in a circular arc. The bookcases are narrow from front to back, and for purposes of the problem we idealize them as straight line segments. What is the minimum number of swings required to reverse the two bookcases?'

489. Put the Cherry in the Glass This diagram represents a cocktail glass, composed of four matches, and a cherry.

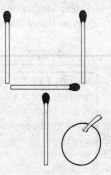

By moving only two matches, place the cherry into the glass.

490. The Bridge of Matches In Cambridge, over the River Cam behind Queens' College, is a bridge which, it is claimed, was originally designed by Sir Isaac Newton without the use of any joints or pins.

This is your chance to imitate the great man. Can you make a bridge, using no glue or other adhesive materials or devices, using twenty-two kitchen matches?

491. The Mystic Square This square has occult properties. For example, careful study of the square will reveal the missing symbol which should go into the empty cell. What is it?

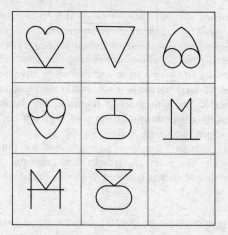

492. Cannonball Quiz   Charles Hutton, Professor at the Royal Military Academy and translator of Ozanam's *Mathematical Recreations*, naturally instructed his students in methods of calculating the amount of enemy ordnance. In particular, he explained how to calculate the number of cannonballs in a pile from the observed number along one edge.

No doubt you can solve the puzzle without Hutton's instruction. On this occasion you observe that the enemy have assembled their cannonballs in one square pyramid, and you are about to raise your telescope to count the number of cannonballs along the bottom edge, when you see an enemy soldier walk up with one more cannonball. The Master of Ordnance appears to remonstrate with him, and then to take the original square pyramid apart and build a new, triangular pyramid, using all the original balls, and the extra one.

How many cannonballs do the enemy possess?

493. Squares into Squares?   Is it possible to assemble a number of different geometrical squares, to make a larger square, leaving no spaces? See how close you can get in assembling squares of the following edge-lengths into a larger square: 99, 78, 77, 57, 43, 41, 34, 25, 21, 16 and 9.

494. Book Words   The Ruritanian National Library contains more books than any single book on its shelves contains words. Also, no two of its books contain the same number of words.

Can you say how many words are in one of its books?

495. Archimedes' Breath   At a rough and ready estimate, what is the probability that you are, at this very moment, breathing in at least one molecule of air that was once breathed by the great mathematician Archimedes?

496. Median Mystery   This is a bit of elementary geometry. A′ and B′ are the mid-points of two sides, the lines AA′ and BB′ meet at X and X divides each line in the ratio 2:1.

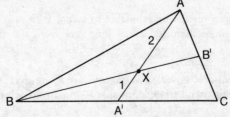

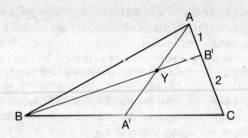

Suppose that B' is moved so that it divides CA in the ratio 2:1, as in the figure. How will the new intersection, Y, divide AA'?

497. Equal Products, Equal Sums It is obvious that 2 × 2 equals 2 + 2. However, 2 and 2 are also two equal numbers. What about different numbers? Which set of different whole numbers has the same product and the same sum?

498. The Island in the Lake The figure shows a small island, on which is a tree, in the middle of a large and deep lake, which is 300 yards across. On the shore is another tree.

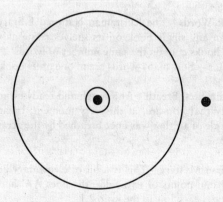

How might a man, who is unable to swim, with only a length of rope rather more than 300 yards long, get from the shore to the island?

499. A Present-able Poser Five pigeons are flying over a field in the form of an equilateral triangle, of side 100 metres. Each pigeon, as it flies, makes a small deposit, as pigeons are wont to do. Explain why at least one pair of these small deposits must be at most 50 metres apart.

500. Five Points on a Lattice Five points are chosen on a square lattice; in other words, five points of intersection are chosen on a square grid. The figure illustrates just one possibility.

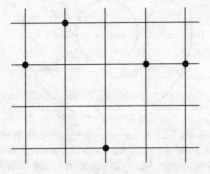

Why is it certain that at least one mid-point of a line joining a pair of the chosen points, is also a lattice point?

501. Six on Five How can six matches be placed on a table so that each of the matches touches all the other five matches?

502. The Last Match More and Less were playing a simple game. They had a pile of twenty-one matches in front of them on the table and they took turns to remove up to, but not more than, three matches. The loser was the person who took the last match.

So far, Less has been less successful than More. Can you recommend a strategy to Less which would make him more successful than More, or at least guarantee that the games were split more or less evenly between them?

503. Squares and Triangles How can eight matches be placed so as to form no less than two perfect squares and four triangles?

504. How Many Triangles With three lines only one triangle can be created, with four lines only four. How many can be created with six straight lines?

505. A Map-colouring Problem It is easy to demonstrate that at least four colours are needed to colour a map so that adjoining countries are differently coloured.

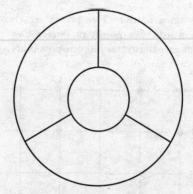

This figure shows four countries, each of which borders the other three. However, only three of the countries are the same shape. Can you remedy this defect by drawing a map of four countries, of identical shape and size, so that each borders the other three?

Langley's Adventitious Angles This tricky problem, named after E. M. Langley, is famous because it is not as simple as it seems.

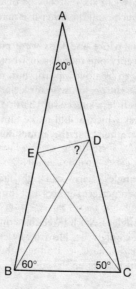

506. ABC is an isosceles triangle, whose vertex angle is 20°. DBC = 60° and ECB = 50°. All you have to do is to find angle BDE.

507. Heads over Tails Lay out eight pennies in the circle, all heads up. Start at any coin of your choice, count four as you touch four coins in succession, and turn over a head. Choose one of the remaining heads, count to four, starting with the chosen coin, and turn over the fourth coin, to show tails up. Repeat until all but one of the coins are tails up. Remember, you must start each time with a head, and the fourth coin must be a head until you turn it tails up.

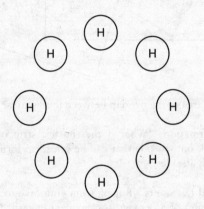

508. Imperfect Products 'Prove that the product of four consecutive positive integers cannot be a perfect square.'

509. The Maximal Product 'What is the largest number which can be obtained as the product of positive integers which add up to 100?'

510. The Programmer's Shirts 'A neat computer programmer wears a clean shirt every day. If he drops off his laundry and picks up the previous week's load every Monday night, how many shirts must he own to keep him going?'

511. The Overlapping Squares Two squares are shown overleaf, the larger being 10 inches square and the smaller being rather smaller. A vertex of the second square lies at the centre of the first square, and the centre of the second square lies on the right-hand edge of the first square, one quarter of the way up the edge.

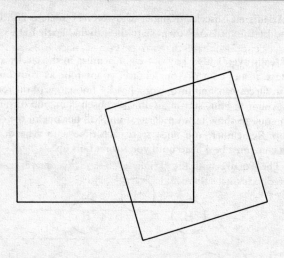

What is the area of the overlap between the squares?

512. Cube Formation 'What is the shortest strip of paper 1 inch wide and black on one side that can be folded to form a 1 inch cube that is black on all sides?'

513. Delightful Discounts Buying from your favourite store you are offered a discount of 5 per cent for payment in cash, 10 per cent as a long-standing customer, and 20 per cent because it is sale time. In what order should you take these discounts in order to pay as little as possible for your purchase?

514. Fold and Fold Again Taking a large rectangular piece of thin paper in your hands, you fold it in half once, and then in half again. Repeating the same action, you fold it fifty times, each time in half. After a few folds it is noticeably thicker. How thick is it after fifty folds?

To be more precise, suppose that the original sheet is one-tenth of a millimetre in thickness.

515. Pandigital Difference I think of a number which contains all the digits 1 to 9, exactly once. I reverse it, so that the first digit becomes the last, and find the difference between the numbers. The answer also contains the digits 0 to 9. What number did I think of?

516. Pandigital Square I think of a number which, curiously, remains the same when I turn it upside down. When I square it, the product contains all the digits from 0 to 9 exactly once each. What number did I think of?

517. Folding a Square from a Rectangle 'You are given a rectangle of paper . . . the dimensions of which are unknown. You are required to determine the side of a square which has an area equal to the rectangle by merely folding the paper three times.'

518. The Square and the Triangle These five pieces can be assembled to form a square and a triangle. How?

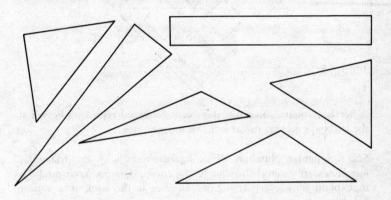

519. How Many Friends? At a party (meaning any gathering of more than two people), at least two people will have the same number of friends present – true or false?

520. Friends and Strangers At a small dinner party, which for the purposes of this problem means a gathering of exactly six people, there will always be either three people who are mutual friends, or three guests who are mutual strangers. True or false?

521. Up and Down the Garden Path Lady Merchant's garden consists of square plots of flowers surrounded by low box-hedges, with paths between the plots.

Lady Merchant enters at the gate to the left and walks along the paths to the summer house at the right-hand corner, every evening taking, as far as possible, a different route.

For how many successive days can she avoid repeating herself if she is always moving towards the summer house?

522. Triangular Numbers This figure shows why the triangular numbers were given their name by the ancient Greeks. Your puzzle is to explain why every triangular number is the sum of a square number and two other triangular numbers.

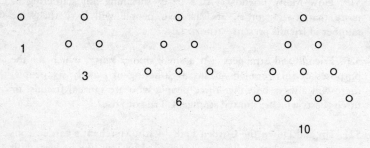

A further question: why is *one more than* every alternate triangular number also the sum of a square and two triangular numbers?

523. Tick and Cross   There are twelve pentominoes, each composed of five identical squares edge to complete edge. You will notice that two of them resemble a V and an X, which we have called a Tick and a Cross. How can an enlarged copy of each of these two figures, three times as wide and three times as tall, be assembled each using nine of the pentominoes?

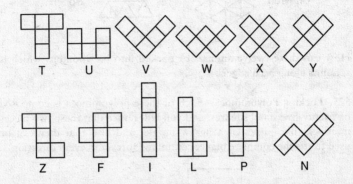

524. Animals in the Cage   Just as there are twelve distinct pentominoes, so there are also, coincidentally, just twelve little 'animals' that can be composed of six equilateral triangles fixed edge to complete edge.

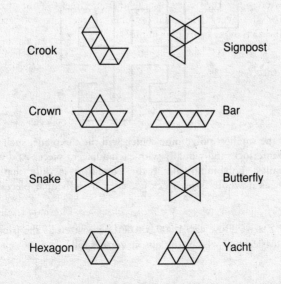

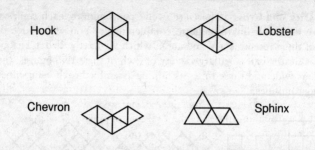

Hook   Lobster

Chevron   Sphinx

How can these twelve animals be packed into the rhombus which is six units along each edge?

525. Locking Polyominoes   Each of these polyominoes is composed of twenty-five small squares, and you will notice that these two pieces interlock, like pieces of a jigsaw puzzle, and also that they can be used to tile the complete plane, continuing forever in every direction.

What is the smallest polyomino which will tile the plane, such that each piece interlocks individually with each adjacent piece? And what is the smallest polyomino tile if the condition is only that the tessellation as a whole is interlocking, even if individual pieces are not?

526. Zero Zeros   How can 1,000,000,000 be written as the product of two factors each of which contains no zeros at all?

527. Two Children 'I have two children. They aren't both boys. What is the probability that both children are girls?'

Now suppose that I have two children of whom the elder is a boy. What is the probability that both are boys?

528. Pearls and Jars 'Mrs Tabako has fifty natural pearls, fifty cultured pearls and two Ming jars. If she uses all the pearls, how should she distribute them in the two jars in such a way that when Mr Tabako enters the room and picks one pearl out of either jar at random he will have the best possible chance of picking a cultured pearl?'

529. The Chord in a Circle If a chord is drawn at random in a circle, what is the probability that it will be longer than the side of this equilateral triangle, inscribed in the circle?

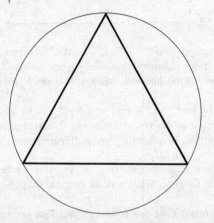

530. Reptile Repeat It is quite easy – well, fairly easy – to cut this rectangle-with-a-corner-misplaced into two identical pieces, as the dotted lines show.

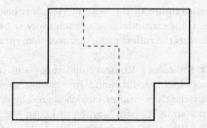

Can you, however, cut it into *three* identical pieces? Three whole pieces, that is; none of the pieces may be made up of smaller parts.

531. The Sphinx This shape, named for obvious reasons, can be cut into four whole pieces, all identical in shape and all the same shape as the original Sphinx. Curiously, the extra lines drawn total one half of the perimeter of the original figure in length. How is it done?

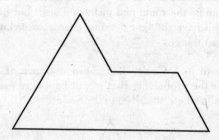

532. Reproducing Reptile 'This is a reptile,' explained Peter, 'it can be made up from identical smaller copies of itself. In fact,' he continued, 'this shape has four sides and is made of four copies of itself.'

'Really,' said Jane, 'that's very obvious,' and to show her disgust she tore it into two halves, exclaiming, 'Each of these halves is a much better reptile, because they are not symmetrical, and it is much harder to see the answer.'

Sure enough, each half of the original shape was a reptile, divisible into four copies of itself. What was the original shape?

533. Simple Sums Take any four-digit number, arrange the digits in ascending and descending order to form two numbers, and subtract the smaller from the larger. Repeat the same process with the answer. What is the result – eventually?

534. Before the Invention of the Wheel A slab is being transported on three rollers. If the circumference of each roller is 1 metre, how far will the slab move as the rollers make one complete rotation?

535. The Most Ridiculous Route A postman with time to spare, made a point of finishing his round by walking as far as possible while visiting his last ten houses, which were equally spaced, 100 metres apart on a straight road. Starting at house No. 1 he delivered its mail and then walked to 10, then back to 2, then all the way to 9, and so on, zig-zagging up and down the road, and ending up at No. 6, where he was always offered a cup of tea and a bun, after walking $100 \times (9 + 8 + 7 + 6 + 5 + 4 + 3 + 2 + 1) = 4500$ metres.

One day, however, it occurred to him that he might do 'worse' than start at No. 1 and he planned an even longer route, which still, however, ended up at No. 6. What was it?

536. Bemusing Bolts   Hold two identical bolts against each other, as in the figure, and rotate them around each other, as if you were 'twiddling your thumbs'.

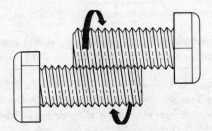

Will the bolts move apart or move closer together?

537. Packing Triangles   I have two triangles, one larger than the other. The longest, middle and shortest sides of the smaller are shorter than the longest, middle and shortest sides of the larger, respectively.

Can I be certain that the smaller triangle can actually be placed inside the larger, without overlapping its edges?

538. Choosing in the Dark   Miss Golightly is getting dressed in a hurry, but the light in her closet has gone out. How many stockings must she take from the stocking drawer, to ensure that she has a pair of the same colour, if there are stockings of seven different colours in the drawer?

539. Conway's Solitaire Army   On an infinite square grid, an army of men stand behind a starting line, waiting to move forward. Every move consists of one man jumping over an adjacent man into the empty square beyond, just as in solitaire. (The jumps may be made horizontally, vertically or diagonally.)

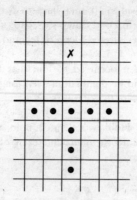

The figure, which is merely illustrative, shows how an army of only eight men can send one man to the third rank beyond the starting line.

Your problem, as General, is to decide just how far the army is able to march. You are allowed, of course, to choose the size of the army and to dispose your men in any manner you choose.

540. Points in a Square   This is a square lattice. The points are all at the vertices of identical squares, and you have to imagine, of course, that the points are infinitely small. This is an essential point – pardon the pun – because the problem is to decide whether it is possible to draw a square on the lattice which contains exactly seventeen lattice points in its interior and no lattice points on its perimeter.

More generally, is there always a square which contains exactly N lattice points in its interior, where N is any integer you choose?

541. The Squirrel and the Hunter   A hunter sees a squirrel in a tree, and walks towards it. As he does so, the squirrel disappears round the far side of the trunk, and as the hunter circles the tree the squirrel keeps out of sight on the other side, circling also. As this curious chase continues there is no doubt that they are both circling the tree, but, are they circling each other?

William James, the famous psychologist, posed this problem in his book *Pragmatism*. What is your pragmatic response?

542. Farthing Fiddle   One of the advantages of the decimal system of coinage is that any sum of pence can instantly be converted to pounds by inserting a decimal point. In the days of pounds, shillings and pence this was not possible – in general. However, there was one five-figure quantity of farthings which could be converted into £.s.d. by simply inserting two strokes of the pen.

What was it? You may recall that £1 = 20 shillings = 240 pence, and a farthing is a quarter of one penny.

543. To Knot or not to be Knotted?   Glancing at this loop of string on a table, too quickly to notice which bits go over which other bits, you idly ask yourself whether it is likely to be knotted? What is the answer?

544. Cooked Turkey 'An old invoice showed that seventy-two turkeys had been purchased for "–67.9–". The first and last digits were illegible.'

How much did each turkey cost?

545. The Chauffeur Problem 'Mr Smith, a commuter, is picked up each day at the train station at exactly 5 o'clock. One day he arrived unannounced on the 4 o'clock train and began to walk home. Eventually he met the chauffeur driving to the station to get him. The chauffeur drove him the rest of the way home, getting him there 20 minutes earlier than usual.

'On another day, Mr Smith arrived unexpectedly on the 4.30 train, and again began walking home. Again he met the chauffeur and rode the rest of the way with him. How much ahead of usual were they this time?'

546. An Express Problem An express train takes 3 seconds to enter a tunnel which is 1 km long. If it is travelling at 120 km an hour, how long will it take to pass completely through the tunnel?

547. The Lost Paddle 'A man went upstream from his dock in a motorboat. As he passed under a bridge one mile from the dock his emergency paddle fell overboard, a loss which he did not discover until 10 minutes later, whereupon he went back downstream to retrieve his paddle, and caught up to it directly opposite his dock. If he travelled at constant water speed and lost no appreciable time turning round, what was the rate of the current of the river?'

548. Spot the Blunder Puzzle-solvers must be wide awake to solutions which seem to be solid as a rock but actually contain large holes. Here is a puzzle and the solution as published in a book which shall be nameless. Spot the boob!

A spy is watching the Pentagon, which as you know is a large building in the form of a regular pentagon, from a distance with powerful binoculars. What is the chance that he or she can see three sides of it?

This is the offered solution: Imagine another spy at an equal distance away, exactly opposite to the first spy. If one spy can see two sides, the other will see three. Since it is equally likely that the spy will be at either spot, the probability is one-half.

549. The Burning Candles 'On Christmas Eve two candles, one of which was one inch longer than the other, were lighted. The longer one was lighted at 4.30 and the shorter one at 6.00. At 8.30 they were both the same length. The longer one burned out at 10.30, and the shorter one at 10.00. How long was each candle originally?'

550. The Heavy Boxes 'Five equal cubical boxes, each with an A on its top side, stand together as in the first figure.

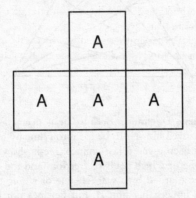

'The boxes are to be brought into line, but they are so heavy that they can be moved only by tipping them over about an edge. With these conditions, it proves to be impossible to bring them into line with all the A's the same way up, and the arrangement finally achieved has the plan view shown in the second figure. Which box was originally in the middle?'

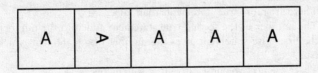

551. Passing Trains A man standing on a platform notes that a train going in one direction takes 3 seconds to pass him, and a train of the same length in the other direction takes 4 seconds. How long did it take for them to pass each other?

552. Triangles in a Triangle   How many triangles can be counted in this figure?

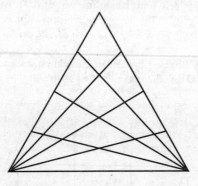

553. The Prisoner's Dilemma   Because he is deemed to be a foolish man who has allowed himself to be led into crime by his companions, the prisoner has been given a last chance. He is shown two doors in the courtyard, one of which leads to freedom and the other to a long sentence. Each is guarded by a warder, one of whom always lies and one of whom is impeccably honest, but he does not know which is which.

He is allowed one question, to be put to one of the warders. How can he discover which is the door to freedom?

554. Short-list

'1. The number of the first true statement here added to the number of the second false statement gives the number of a statement which is true.

2. There are more true statements than false.

3. The number of the second true statement added to the number of the first false statement gives the number of a statement which is true.

4. There are no two consecutive true statements.

5. There are at most three false statements.

6. If this puzzle consisted of statements 1 to 5 only, then the answer to the following question would still be the same.

Which statements are true?'

555. The Crossed Cylinders   Two identical cylinders are placed so that their axes cross at right-angles and their common volume, has four identical curved surfaces.

How can the volume of this common solid be calculated without the use of any calculus?

556. Concyclic Points   'Five paper rectangles – one with a corner torn off – and seven paper disks have been tossed on a table. They lie as shown in the figure. Each corner of a rectangle and each spot where edges intersect makes a point. The problem is to find three sets of four concyclic points: four points that can be shown to lie on a circle. For example, the corners of rectangle R are such a set, because the corners of any rectangle lie in a circle. What are the other two sets?'

557. Hot Cross Buns 'The hot cross bun man cried:

> Hot cross buns, hot cross buns,
> One a penny, two a penny, hot cross buns,
> If your daughters don't like them
> Give them to your sons!
> Two a penny, three a penny, hot cross buns,
> I had as many daughters as I have sons
> So I gave them seven pennies
> To buy their hot cross buns.

How many children were there if they were all treated alike and if there was only one way in which to purchase the buns?'

558. White to Play This looks like the position after White has made a rather unusual first move, yet it is, in fact, the position after Black has just played.

What is the smallest number of moves that could have been played, in order to reach this position with White to play?

559. The Unwound Clock 'I have no watch, but I have an excellent clock, which I occasionally forget to wind. Once when this happened I went to the house of a friend, passed the evening in listening to a radio concert programme, and went back and set my clock. How could I do this without knowing beforehand the length of the trip?'

560. Tom's House 'John is trying to find out where Tom lives, and all he knows is that it is in a street where the houses are numbered from 8 to 100 (inclusive). John asks, "Is it greater than 50?" and Tom answers, but lies. John then asks, "Is the number a multiple of 4?" Again Tom answers, and again he lies. Then John says, "Is it a perfect square?" Tom answers and this time he tells the truth. Finally John asks "Is the first digit 3?" After Tom has replied (truthfully or not we do not know!) John tells him the number. He is wrong! What was the number of Tom's house?'

561. The Same Sister Is it possible for two men who are completely unrelated to each other, to have the same sister?

562. C is Silent 'On the Island of Imperfection there are three tribes, the Pukkas, who always tell the truth, the Wotta-Woppas, who never tell the truth, and the Shilli-Shallas, who make statements which are alternately true and false, or false and true.

'As the reader can imagine, it is always the most important part of life on the island to discover to which tribe people belong. On a recent visit I was doing some work on this with three inhabitants whom I shall call A, B and C. They have got into the habit lately of going around in threes, one from each tribe, and I am glad to say that these three were no exception.

'C did not make my self-appointed job as a detective any easier by being silent, but the other two spoke as follows:

A: "C is a Pukka."
B: "A is a Pukka."

'Find the tribes to which A, B and C belong.'

563. A cylinder can be 'squared' with the use of only ten square pieces. How can squares of edges 30, 27, 25, 17, 15, 13, 11, 8, 3 and 2 be fitted together to fill the space between two parallel lines, in such a way that when the opposite edges are joined, it forms a 'squared' cylinder?

564. Speaking of Bets ' "The three of us made some bets.

1. First, *A* won from *B* as much as *A* had originally.
2. Next, *B* won from *C* as much as *B* then had left.
3. Finally, *C* won from *A* as much as *C* then had left.
4. We ended up having equal amounts of money.
5. I began with 50 cents."

'Which of the three – *A*, *B* or *C* – is the speaker?'

565. The Murderess 'Three women, named Anna, Babs and Cora, were questioned about the murder of Dana. One of the three women committed the murder, the second was an accomplice in the murder, and the third was innocent of any involvement in the murder.

'Each of the following three statements was made by one of the three women:

1. Anna is not the accomplice.
2. Babs is not the murderess.
3. Cora is not the innocent one.

I. Each statement refers to a woman other than the speaker.
II. The innocent woman made at least one of these statements.
III. Only the innocent woman told the truth.

Which one of the three women was the murderess?'

566. Six Gs In the multiplication problem below, each letter represents a different digit:

$$A\,B\,C\,D\,E$$
$$\times\ F$$
$$\overline{G\,G\,G\,G\,G\,G}$$

Which of the ten digits does *G* represent?

567. The Wheels of Commerce ' "How's the motor business?" asked Bob, glancing at the menu.

'Ben owns a used car lot. His cars are good; his prices are right; his guarantee means just what it says. Other dealers come and go, but Ben keeps right on selling. "Not too bright around Christmas," he replied, "but sales have picked up again."

' "That's dandy!" commented Bob. "I was talking to Stan Logan down on Wardie and Myrtle yesterday. He's hardly sold a car this year."

'Ben smiled. "A lot of them are having a tough time," he said, "but maybe I'm lucky. We've done well so far this month – each week more sales than the previous week."

'"What's that in actual numbers?" asked Bob, who's a great one for facts.

'"I'm not sure about the last few days," replied Ben, "but we sold fifty-six cars the first three weeks. And here's something to amuse yourself with." He thought a moment. "The difference between the numbers we sold in the first and second weeks, multiplied by the difference between the second and third weeks, comes to the same as the number we sold the first week."

'The shapely waitress leaned over his friend just then to take their order, and Bob rather lost interest in car sales. But how many cars would you say Ben sold in the third week?'

568. Professor Mesozoic, the famous geologist, had a problem. She had lost her notes on the samples of sedimentary rock that she had collected, and she no longer knew in what orientation they had been found. Then her assistant, Slatebed, had a bright idea. Within the rock were numerous tiny specks of a mineral, which it was reasonable to suppose had been randomly distributed in the material when it was laid down on the bed of some ancient lake. Subsequently, as the sediment was vertically compressed into rock, they would have been forced together in that vertical direction, and examination of their present distribution would show what that direction had been!

He explained his idea enthusiastically, but Professor Mesozoic thought for a moment, and then rejected his idea. Why?

The Solutions

1. $7 + 49 + 343 + 2401 + 16{,}807 = 19{,}607$.

2. $7 + 49 + 343 + 2401 + 16{,}807 + 117{,}649 = 137{,}256$.
[Boyer, 1985, p. 210]

3. One! All the others were coming *from* St Ives!
[Midonick, 1965, quoting *Every Child's Mother Goose*, introduction by Carolyn Wells, New York, 1918]

4. Each of the five solutions has nine terms. $1 = 1/3 + 1/5 + 1/7 + 1/9 + 1/11 + 1/15 + 1/35 + 1/45 + 1/231$ has the smallest larger denominator, 231.
[Gardner, 1978a]

5. The smallest value of the denominator is 25: the greedy algorithm gives $3/25 = 1/9 + 1/113 + 1/25425$, but $3/25$ also equals $1/10 + 1/50$.
[Gardner, 1978a]

6. $1/2 = 1/2^2 + 1/3^2 + 1/4^2 + 1/5^2 + 1/7^2 + 1/12^2 + 1/15^2 + 1/20^2 + 1/28^2 + 1/35^2$.
[Szurek, 1987, p. 391]

7. 9.
[Peet, 1923, p. 63]

8. $13\frac{1}{23}$.
[Peet, 1923, p. 65]

9. $16 + 1/56 + 1/679 + 1/776$. This and similar problems were solved by the *rule of false position*. An answer which was judged to

be roughly correct was chosen, and then adjusted by multiplying by a suitable factor.

In this case the scribe guessed 16 immediately, and got the wrong but close answer $36 + 2/3 + 1/4 + 1/28$, which falls short of 37 by $2/42$. Next the scribe calculates that $1 + 2/3 + 1/2 + 1/7$ is $97/42$ so that the multiplier required is $2/97$, which the scribe could read off from the earlier table in the Rhind papyrus of fractions $2/n$, without any further calculation. It is $1/56 + 1/679 + 1/776$.

[Peet, 1923, p. 69]

10. $1\frac{2}{3} + 10\frac{2}{3}\frac{1}{6} + 20 + 29\frac{1}{6} + 38\frac{1}{3} = 100$. This problem was also solved by *false position*. The scribe first artificially constructs the series $1 + 6\frac{1}{2} + 12 + 17\frac{1}{2} + 23 = 60$, which has the property that the first two terms sum to one-seventh of the last three. Each term is then multiplied by $1\frac{2}{3}$ to change the 60 into 100.

[Peet, 1923, p. 78]

11. $6^2 + 8^2 = 100$.
[Gillings, 1972, p. 161]

12. The difference between each successive share is $37,55$ or $37\frac{11}{12}$.
[Neugebauer and Sachs, 1945, p. 53]

13. $1000 = 10^2 + 30^2$.
[Eves, 1976, p. 46]

14. The distance is the third side of a right-angled triangle with hypotenuse $0,30$ and one leg $0,30 - 0,6 = 0,24$. The third leg is length $0,18$.

[Eves, 1976, p. 46]

15. Given that a, b and c are integers, such that $a^2 + b^2 = c^2$, either a or b is even; suppose that a is even.

Then there are integers p and q such that $a = 2pq$, $b = p^2 - q^2$ and $c = p^2 + q^2$.

The evidence that the Babylonians used this formula is simple: the values of p and q which fit the numbers on Plimpton 322 are all so-called 'regular' numbers whose factors are powers of only 2, 3, and 5 or products of such powers.

The ratio $c/a = \frac{1}{2}(p/q + q/p)$. The Babylonians could now find suitable values of p and q by referring to the standard reciprocal tables which they used for multiplication anyway. We lack such tables, so to make this equal to approximately 1.54, put $p/q = t$ and solve the quadratic $\frac{1}{2}(t + 1/t) = 1.54$:

$$t^2 + 1 = 3.08t$$
$$t \doteq 2.711 \text{ or } 0.369$$

We discard 0.369, as we want $p > q$:

Take $\dfrac{p}{q} = \dfrac{27}{10}$ as a rough approximation, so that

$$p = 27, q = 10$$
Then $\quad a = 2pq = 540$
$$c = p^2 + q^2 = 829$$
and $\quad \dfrac{c}{a} = 1.535$

The approximate ratio 1.54 in the problem was taken from the figures 3541 and 2291 in Plimpton.

[Neugebauer and Sachs, 1945, pp. 38–41; Eves, 1976, p. 37]

16. Let the letters X, Y, Z and T denote the numbers of white, black, dappled and yellow bulls respectively, and x, y, z and t denote the number of white, black, dappled and yellow cows, respectively. Then the conditions of the problem give seven equations in these eight unknowns:

(1) $X - T = 5/6\ Y$ (4) $x = 7/12\ (Y + y)$

(2) $Y - T = 9/20\ Z$ (5) $y = 9/20\ (Z + z)$

(3) $Z - T = 13/42\ X$ (6) $z = 11/30\ (T + t)$

(7) $t = 13/42\ (X + x)$

From the first three equations, X, Y and Z can be found in terms of T:

$$X = 742/297\ T \qquad Y = 178/99\ T \qquad Z = 1580/891\ T$$

Since 891 and 1580 possess no common factors, T must be some whole multiple – let us say G – of 891. Consequently,

$$X = 2226G \qquad Y = 1602G \qquad Z = 1580G \qquad T = 891G$$

If these values are substituted into equations (4), (5), (6) and (7), the following equations are obtained:

$$12x - 7y = 11214G \qquad 20y - 9z = 14220G$$

$$30z - 11t = 9801G \qquad 42t - 13x = 28938G$$

These equations are solved for the four unknowns x, y, z and t and we obtain:

$$4657x = 7206360G \qquad 4657y = 4893246G$$

$$4657z = 3515820G \qquad 4657t = 5439213G$$

in which the number 4657 is prime. Since it divides none of the coefficients on the right, 4657 must divide G. Taking the simplest case, let $G = 4657$, we obtain as the smallest solution:

white bulls	10,366,482	white cows	7,206,360
black bulls	7,640,514	black cows	4,893,246
dappled bulls	7,358,060	dappled cows	3,515,820
yellow bulls	4,149,387	yellow cows	5,439,213

[Archimedes' cattle problem, as taken from T. L. Heath, *The Works of Archimedes with the Method of Archimedes*, Dover, n.d., p. 319. This solution follows Dorrie, 1965]

17.

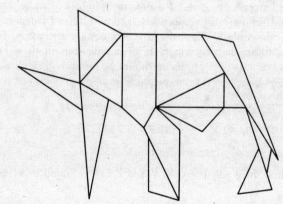

18. Suppose that Mary aims for point P on the river bank. Reflect Mary's original position in the line of the river bank. Then the distance SPT equals the distance S′PT and the latter will be a minimum when S′PT is a straight line. It follows that P is the point such that SP and PT make the same angle with the line of the river.

Heron used exactly the same argument by reflection to conclude that when light is reflected, the angles of incidence and reflection are equal. This is one of the earliest solutions to an extremal problem. As

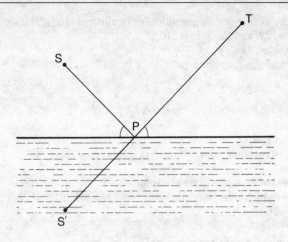

one Greek commentator remarked on Heron's solution, expressing a view which has haunted and inspired scientists ever since, 'for Nature does nothing in vain nor labours in vain.'

19. 'My right eye fills 1/8 jar in 6 hours [taking a day to be 24 hours, where the Greeks might have taken it to be 12], and my left eye fills 1/12 in 6 hours, and my foot 1/16. Thus all four fill the jar 1 + 1/8 + 1/12 + 1/16 = $1\frac{13}{48}$ times in 6 hours. So the jar will be filled once in 6 × 48/61 hours, or 47 minutes and 13 seconds, approximately.'
    [Sandford, 1930, p. 216]

20. A, B and C do the whole work in 10, 15 and 30 days, respectively.

21. Suppose that the given ratio is n rather than 3. Then, if u,v and x,y are the sides of two such rectangles, the equations can be written:

$$u + v = n(x + y) \qquad xy = nuv$$

and Heron's solution runs parallel to the general solution:

$$x = 2n^3 - 1 \qquad y = 2n^3$$
$$u = n(4n^3 - 2) \qquad v = n$$

which leads Heron to his solution: the rectangles are 53 × 54 and 318 × 3.
    [Thomas, 1980, p. 505]

22. There is no very simple solution to this problem. The sides are 20, 21 and 29, and the area is 210.

[Thomas, 1980, pp. 507–8]

23. 20.

24. The first number is 98, the second, 94.

25. 9, 7, 4 and 11.

[The quotation from Xylander is from Ore, 1948, p. 195]

26. 6, 4 and 5 is the simplest solution, but it is the ratios of the numbers which are important, so, for example, 12, 8 and 10 is another solution.

27. Diophantos's answer is 1, 7 and 9.

This is his solution, which illustrates very well his methods, which tend to simplify the problem and produce one or a few solutions. It was left to later Indian mathematicians to find more general solutions.

'Take a square and subtract part of it for the third number; let $x^2 + 6x + 9$ be one of the sums, and 9 the third number.

'Therefore product of the first and second = $x^2 + 6x$; let first = x, and that second = $x + 6$.

'By the two remaining conditions, $10x + 54$ and $10x + 6$ are both squares.

'Therefore we have to find two squares differing by 48; this is easy and can be done in an infinite number of ways.

'The squares 16, 64 satisfy the condition. Equating these squares to the respective expressions, we obtain $x = 1$ and the numbers are 1, 7, 9.'

28. This is Diophantos's solution:

'Let the sum of all three be $x^2 + 2x + 1$, sum of first and second x^2, and therefore the third $2x + 1$; let sum of second and third be $(x - 1)^2$.

'Therefore the first = $4x$, and the second = $x^2 - 4x$.

'But first + third = square, that is, $6x + 1$ = square = 121, say.

'Therefore $x = 20$ and the numbers are 80, 320, 41.'

29. 'Let x = the whole number of measures; therefore $x^2 - 60$ was the price paid, which is a square = $(x - m)^2$, say.

Now $\frac{1}{5}$ of the price of the five-drachma measures + $\frac{1}{8}$ of the price of the eight-drachma measures = x;

so that $x^2 - 60$, the total price, has to be divided into two parts such that $\frac{1}{5}$ of one + $\frac{1}{8}$ of the other = x.

We cannot have a real solution of this unless $x > \frac{1}{8}(x^2 - 60)$ and $< \frac{1}{5}(x^2 - 60)$.

Therefore $5x < x^2 - 60 < 8x$.

(1) Since $x^2 > 5x + 60$,

$x^2 = 5x$ + a number greater than 60,

whence x is *not less than* 11.

(2) $x^2 < 8x + 60$

or $x^2 = 8x$ + some number less than 60,

whence x is *not greater than* 12.

Therefore $11 < x < 12$.

Now (from above) $x = (m^2 + 60)/2m$;

therefore $22m < m^2 + 60 < 24m$.

Thus (1) $22m = m^2$ + (some number less than 60),

and therefore m is *not less than* 19.

(2) $24m = m^2$ + (some number greater than 60),

and therefore m is *less than* 21.

Hence we put $m = 20$, and

$x^2 - 60 = (x - 20)^2$,

so that $x = 11\frac{1}{2}$, $x^2 = 132\frac{1}{4}$, and $x^2 - 60 = 72\frac{1}{4}$.

Thus we have to divide $72\frac{1}{4}$ into two parts such that $\frac{1}{5}$ of one part *plus* $\frac{1}{8}$ of the other = $11\frac{1}{2}$.

Let the first part be $5z$.

Therefore $\frac{1}{8}$ (second part) = $11\frac{1}{2} - z$,

or second part = $92 - 8z$;

therefore $5z + 92 - 8z = 72\frac{1}{4}$;

and $z = 79/12$.

Therefore the number of five-drachma measures = 79/12.

Therefore the number of eight-drachma measures = 59/12.'

    [Midonick, 1968, pp. 48-9]

30. The string must form a semi-circle. Imagine that it takes the form in the figure, and reflect the shape in the shore-line. Then the entire closed curve will be the curve that encloses the largest area, for double the length of string. This is a circle, a fact which follows from the theorem that the polygon with a given number of sides, with

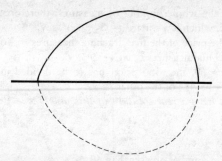

maximum area, is a regular polygon, if the number of sides is then allowed to tend to infinity.

31. The area is a maximum when the ends of the rods lie on a circle. This conclusion is suggested by the thought that if the quadrilateral is adjusted so that its vertices do lie on a circle, which is certainly possible, and the four arcs of the circle are then also hinged at the vertices of the quadrilateral, and the figure moved, then the area surrounded by the four circular arcs cannot be a maximum since they no longer form a circle: yet the areas between the arcs and the sides of the quadrilateral have not changed – only the interior area of the quadrilateral changes when the figure is moved about its hinges.

The area can be calculated by a formula that was discovered by Brahmagupta but also apparently known to Archimedes. If half the sum of the sides, a, b, c and d, is s, then the area is given by

$$A = \sqrt{(s - a)(s - b)(s - c)(s - d)}$$

(If one of the sides has zero length, then the quadrilateral becomes a triangle, which is automatically inscribable in a circle, and this formula becomes Heron's formula $\sqrt{s(s - a)(s - b)(s - c)}$ for the area of a triangle with sides a, b and c; s is half the of the sides.)

32. Reflect the original figure in both walls, and then reflect a third time, to get this complete figure.

The area enclosed by the screen will be a maximum when the area of the entire octagon is a maximum, and this will be so when it is a regular octagon. So the screen must be placed so that it meets the walls at two angles of $67\frac{1}{2}°$ each.

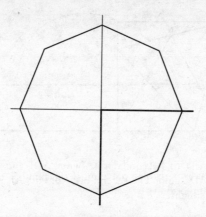

33. Reflect the isosceles triangle in its third, variable, side, to form a rhombus. The area of the rhombus will be a maximum when it is a square, and so the area of the isosceles triangle is a maximum when the angle between its equal sides is a right-angle.

34. $5\frac{1}{7}$ hours are past and $6\frac{6}{7}$ remain.

35. He was a boy for fourteen years, a youth for seven; at 33 he married, and at 38 he had a son born to him who died at the age of 42. The father survived him for four years, dying at the age of 84.

36. $577\frac{7}{9}$ and $422\frac{2}{9}$.
   [Problems 34–6 are from *The Greek Anthology*, 1941]

37. If the estate remaining after payment of the legacy is divided into twenty parts, the husband receives five, the son six and each daughter three. The stranger receives 15/56, so Al-Khwarizmi divides the whole estate into $20 \times 56 = 1120$ parts. The stranger receives 300, the husband 205, the son 246, and each daughter 123.

38. Abul Wafa gave five different solutions. Here are three of them.
   Let one vertex of the equilateral triangle be at D. Construct N so that ABN is equilateral. Mark F on AB so that AB = BF, and draw an arc cutting ABF so that FN = FG. Then G is one of the other vertices, and the last vertex is easily found on BC.

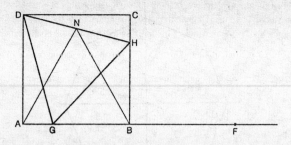

Construct AID to be equilateral. Bisect ADI and then bisect the half towards AD again. The second bisector cuts AB at G, one of the other vertices sought.

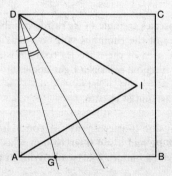

Join B to the mid-point, M, of DC. Draw an arc with centre B and radius BA to cut BM at N. Let DN cut CB at H. Then H is one of the other vertices sought.

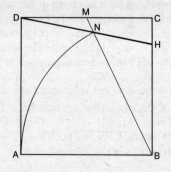

[Berggren, 1986]

39. Abul Wafa's solution bisects two of the squares and places them symmetrically around the third. Joining vertices by the dotted lines, the larger square is found. The four small pieces outside it fit exactly the spaces inside its boundary.

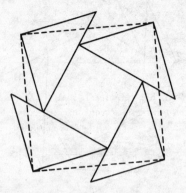

This solution is not as idiosyncratic as it initially appears. If one of the original squares plus a square composed of all four of the 'quarters' are repeated to form a tessellation, then joining the dotted lines is one standard way of dissecting one small square and one large square into a single square. It also works if the larger square is thought of as composed of four quarters.

Note that the simplest dissection of a Greek Cross into a square (see p. 228 below) can also be seen as an example of Abul Wafa's theme.

40. The same solution works if the two larger squares are bisected. Their size is irrelevant.

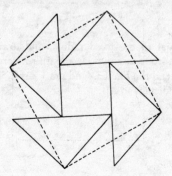

41. Dissect the larger hexagon into six identical triangles, as here, and then arrange them around the smaller hexagon. The dotted lines complete the dissection.

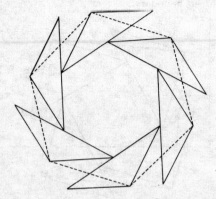

[Wells, 1975]

42. Arrange the three larger triangles round the small similar triangles, like this, and join the vertices as indicated.

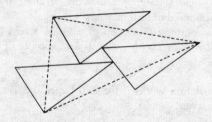

[Wells, 1975]

43. Mark AC equal to the fixed radius, and draw two arcs, with centres A and C, to construct D, the third vertex of equilateral triangle ACD. Extend the line CD and mark off E, so that DE = CD. Then AE is perpendicular to AB.

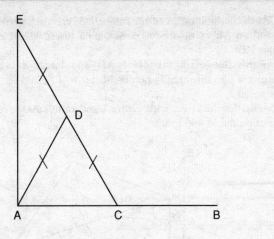

Construct perpendiculars in opposite directions at the endpoints
and B. Mark off as many segments as necessary along each
endicular, using the fixed radius. Joining the first mark on one
endicular to the nth mark on the other, and so on, will divide the
ient AB into $n + 1$ equal parts.

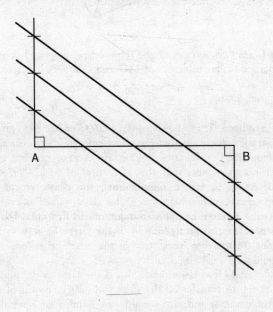

45. 'At the endpoint of A of the radius DA, erect AE perpendicular to AD and on AE mark off AE = AD, then bisect AD at Z and draw the line ZE.

'On this line mark off ZH = AD and bisect ZH at T. Then construct a line through T, perpendicular to EZ, and let it meet DA extended at I.

'Finally, let the circle with centre I and radius AD meet the given circle at points M and L.'

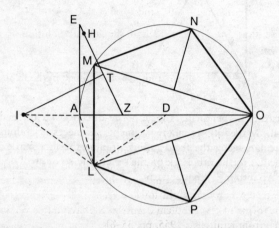

M and L are two vertices of the required pentagon, O is another, and the perpendicular bisectors of MO and LO meet the circle at the other vertices.

[Berggren, 1986]

46. Sissa required $2^{64} - 1 = 18,446,744,073,709,551,615$ grains of wheat. *Cassell's Book of Indoor Amusements, Card Games and Fireside Fun* (1881) calculates, taking an average number of 9216 wheat kernels to a pint, that this is a total of 31,274,997,411,298 bushels of grain, 'a larger amount than the whole world would produce in several years'.

By way of a modern example, the authors further calculate that if one pin were dropped into the hold of the Great Eastern steamship (22,500 tons) in the first week, two in the next and so on, a year's worth would fill 27,924 Great Easterns.

As Kasner and Newman remark, this is the same as the number of moves required to transfer all the rings in Lucas's 'Tower of Hanoi' puzzle (problem 238) and also roughly the number of ancestors that each person alive today had at the start of the Christian era, which happens to be about sixty-four generations ago. The ratio of 2^{64} to the actual population of the earth at that time is therefore a measure of the amount of unintentional interbreeding that has taken place.

[Kasner and Newman, 1949]

47. Let the numbers of men, women and children be m, w and c respectively. Then the problem states that

$$m + w + c = 20 \quad \text{and} \quad 3m + \tfrac{3}{2}w + \tfrac{1}{2}c = 20$$

It follows that $5m + 2w = 20$ and the unique solution is $m = 2$, $w = 5$, and $c = 13$.

48. Fifty soldiers broke down and fifteen remained in the field.
[Mahavira, 1912, p. 112]

49. If b/a and d/c are the original selling prices, then the average price is $\tfrac{1}{2}(b/a + d/c)$. The trusts set the price to be $(b + d)/(a + c)$. Comparing these two expressions, and simplifying, it follows that the trust price will be advantageous only if $a > c$ and $b/a > d/c$, that is, if the original prices are unequal and the denominator of the higher price is greater than that of the lower price.

[The theme of this problem occurs in Mahavira; this version is taken from Kraitchik, 1955, pp. 35–6]

50. After $\tfrac{1}{6}$ and $\tfrac{1}{3}$ have reached the maid-servant and the bed, one half remain. These are halved again and again, six times in all, leaving $1/128 = 1161$. The total number of pearls is therefore the improbable 148,608.

[Mahavira, 1912, p. 73]

51. There are six ways of choosing a single flavour, and $(6 \times 5)/2 = 15$ ways of choosing a pair of flavours. Similarly there are $(6 \times 5 \times 4)/(3 \times 2) = 20$ ways of choosing three flavours, and $(6 \times 5 \times 4 \times 3)/(4 \times 3 \times 2) = 15$ choices of four flavours. This last figure is equal to that for a choice of two flavours because choosing four flavours is the same as choosing two which you will *not* include. By the same reasoning there are six ways of choosing five flavours.

The total of all these answers, including the single way in which all the flavours can be rejected is $2 \times 2 \times 2 \times 2 \times 2 \times 2 = 64$, because each flavour can be either rejected or accepted.

[Mahavira, 1912, p. 150]

52. Let the value of the purse be x, and the wealth of the three merchants, p, q and r. From the equations

$$p + x = 2q + 2r$$
$$q + x = 3p + 3r$$
$$r + x = 5p + 5q$$

it follows that $p:q:r = 1:3:5$, and the solution in smallest integers is that the merchants originally had 1, 3 and 5 in money, and the value of the purse was 15.

53. The number of arrows in a bundle is the sum, as far as necessary, of the series $1 + 6 + 12 + 18 + \ldots$

If eighteen arrows are visible, there are thirty-seven arrows in all.
    [Mahavira, 1912, p. 167]

54. Mahavira does not state the distance between the pillars because this need not be known. The height required is one half of the *harmonic mean* of the given heights, that is, if the heights of the pillars are P and Q, then the required height is

$$\frac{1}{\left(\dfrac{1}{P} + \dfrac{1}{Q}\right)}$$

or

$$\frac{PQ}{(P + Q)}.$$

(The point at which the string touches the ground divides the horizontal distance between the pillars in the ratio of their heights.)

Mahavira also solves the problem in which the strings are attached to the ground at points outside the bases of the pillars.

[Mahavira, 1912, p. 243]

55. The common difference is 22/7.

[Eves, 1976, p. 199; Midonick, 1965, p. 277]

56. From the figure, $x^2 = 15^2 + (45 - x)^2$ from which $x = 25$. Alert readers might spot at once that the triangle is just the enlarged 3–4–5 triangle.

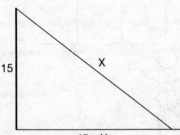

[Eves, 1976, p. 199]

57. $10 \times 9 \times 8 \times 7 \times 6 \times 5 \times 4 \times 3 \times 2 \times 1 = 3,628,800$ possible arrangements of Siva's attributes.

There are $4 \times 3 \times 2 \times 1 = 24$ arrangements of Vishnu's, all of which, according to Midonick, have their own special names.

[Sandford, 1930, p. 198; Midonick, 1965, p. 275]

58. Let the value of each blue gem be b, of each emerald, e, and of each diamond, d. Then,

$$12b + 2e + 2d = 6e + 2b + 2d = 4d + 2e + 2b$$

It follows that the ratios $b:e:d$ are $2:5:10$, and these are the simplest possible integral values for their worth, which cannot be determined more exactly.

59. This is the original *Lo Shu* diagram. The solution is essentially unique, but there are eight possible solutions obtained by merely reflecting and rotating any one of them.

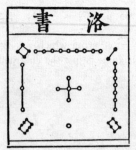

4	9	2
3	5	7
8	1	6

[Needham, 1959, p. 57]

60. Twenty-three pheasants and twelve rabbits.
[Midonick, 1965, p. 183]

61. Seven men and fifty-three articles.
[Midonick, 1965, p. 183; Mikami, 1964, p. 16]

62. Each ox, $1\frac{13}{21}$ tael; each sheep, $\frac{20}{21}$ tael.
[Midonick, 1965, p. 183]

63. $9\frac{1}{4}$, $4\frac{1}{4}$, $2\frac{3}{4}$ measures of grain, respectively.
[Mikami, 1964, p. 18]

64. By Pythagoras, or rather by the Gougu theorem, which was the Chinese name for the theorem of Pythagoras, the water is 12 feet deep. The right-angled triangle formed by the length of the reed excluding the protruding foot, the line from its base to the edge of the pool, and the line from that same point on the edge to the point where the reed breaks the water, is a 12–13–5 triangle.

However, the only requirement of the problem is that the hypotenuse be 1 unit longer than another leg. In general, the formula $(2n + 1)^2 + (2n^2 + 2n)^2 = (2n^2 + 2n + 1)^2$ gives a right-angled triangle with hypotenuse and one leg differing by unity.

[Midonick, 1965, p. 184]

65. 17 feet, by Pythagoras: the figure forms an 8–15–17 triangle. [Midonick, 1965, p. 184]

66. $4\frac{11}{20}$ feet. If the height of the break is x, then $x^2 + 3^2 = (10 - x)^2$.

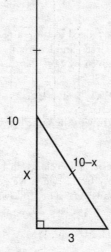

[Midonick, 1965, p. 184]

67. The circle has diameter 6. Liu Hui, the author of the *Sea-Island Arithmetical Classic*, demonstrated this solution by a dissection.

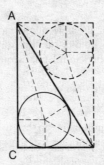

The triangle containing the inscribed circle is doubled to form a rectangle, of dimensions $a \times b$ where a, b and c are the shorter sides and hypotenuse of the original triangle, respectively.

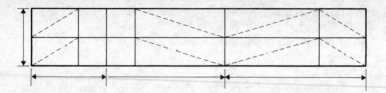

The pieces are then rearranged to form a rectangle of height D, the diameter of the required circle, and length $a + b + c$. It follows that the required diameter is given by $D = 2ab/(a + b + c)$. Here $a = 8$, $b = 15$ and $c = 17$.

[Li Yan and Du Shiran, 1987, p. 71]

68. $2\frac{6}{13}$ days, when both grow to the same height of $4\frac{11}{13}$ feet.
[Mikami, 1964, p. 18]

69. There were sixty guests. The rule given by Sun Tsu is 'Arrange the 65 dishes, and multiply by 12, when we get 780. Divide it by 13, and thus we obtain the answer.' This follows from the fact that if there were x guests, then $\frac{x}{2} + \frac{x}{3} + \frac{x}{4} = 65$.
[Mikami, 1964, pp. 31–2]

70. The numbers in this sequence all leave a remainder 2 when divided by 3: 2, 5, 8, 11, 14, 17, 20, 23, 26, . . . Of these numbers, the sequence of numbers 8, 23, 38, . . ., also leave a remainder of 3 when divided by 5. Out of this sequence, the numbers 23, 128, 233, . . . (where the difference is always $3 \times 5 \times 7 = 105$) also leave a remainder of 2 when divided by 7.

Therefore the smallest possible number of unknown things is 23, but there are in fact an infinite number of solutions to the original problem, i.e. all the numbers in the final sequence.

71. Sixty days, the lowest common multiple of 3, 4 and 5.
[Mikami, 1964, p. 33]

72. If the lengths of the shorter sides are a and b, the side of the square is $ab/(a + b)$, as Liu Hui proved by this beautiful figure, similar to the figure he used to solve problem 67:

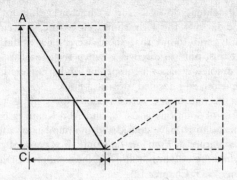

The original diagram is repeated to complete a rectangle $a \times b$, and this is then reassembled to form an equal rectangle of length $a + b$ and height equal to the side of the required square.

[Li Yan and Du Shiran, 1987, p. 70]

73. He catches up $37 - 23 = 14$ miles in 145 miles, so he will catch up 37 miles in $145 \times 37/14 = 383\frac{3}{14}$ miles.

[Mikami, 1964, p. 41]

74. If the numbers of cocks, hens and chicks are c, h and s respectively, then the conditions are:

$$c + h + s = 100$$
$$5c + 3h + \tfrac{1}{3}s = 100$$

which together imply $7c + 4h = 100$.

These equations are indeterminate – there are not enough conditions to fix the values exactly. However, given that only whole numbers of birds were sold, and some were sold of each kind, there are just three solutions:

$$
\begin{array}{lll}
c = 12 & h = 4 & s = 84 \\
c = 8 & h = 11 & s = 81 \\
c = 4 & h = 18 & s = 78
\end{array}
$$

If the number of kinds of articles sold is increased, but with the same information given, then the number of possible solutions also increases dramatically. The Arabic author Abu Kamil, in his *Book of Arithmetical Rarities*, written just before 900 AD, supposed that 100 birds are sold for 100 drachma, the birds being ducks at 2 drachma, hens at 1 drachma, doves 2 for a drachma, ring-doves 3 for 1 drachma and larks 4 for 1 drachma. He wrote, 'I went into [this problem] fully and found that there were 2,696 valid answers. I marvelled at this, only to discover – when I spoke of it – that I was reckoned a simpleton or an incompetent, and strangers looked upon me with suspicion. So I decided to write a book . . .'

This must be one of the better reasons for writing a book. The actual number of solutions is 2,678, and Abu Kamil is probably the first mathematician to have considered the number of solutions of a problem as a feature in itself. At that time, it was usually considered adequate to find one solution, following the example of Diophantos.

[O'Beirne, 1965, Chapter 12]

75. This is Yang Hui's solution.

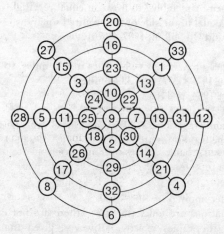

[Needham, 1959, p. 60]

76. 'The ox leaves no trace in the last furrow, because he precedes the plough. However many footprints he makes in the earth as he goes forward, the cultivating plough destroys them all as it follows. Thus no footprint is revealed in the last furrow.'

77. Alcuin answers: 'First there were 250 pigs bought with 100 shillings at the above mentioned rate, for five fifties are 250. On division, each merchant had 125. One sold the poorer quality pigs at three for a shilling; the other the better quality at two for a shilling. The one who sold the poorer pigs received 40 shillings for 120 pigs;

the one who sold the better quality received 60 shillings for 120 pigs. There then remained five of each sort of pig, from which they could make a profit of 4 shillings and 2 pence.'

78. If the servant is not included in the count at each stage, then he would arrive at the first manor having collected no men, so would collect none there, and so on; the total collected would be zero! Therefore the servant must include himself as the first soldier, and the numbers on leaving each manor are 2, 4, 8 . . . , and on leaving the thirtieth manor, $2^{30} = 1,073,741,824$.

79. They are cousins twice over, each having a parent who is sibling to a parent of the other, in two ways.

80. 'To each son will come ten flasks as his portion. But divide them as follows; give the first son the ten half-full flasks; then to the second give five full and five empty flasks, and similarly to the third.'

81. Alcuin's solution, abbreviated, is: I cross with my sister, leave her on the other side, and return. The other two sisters then cross, and my sister brings the boat back. Then the other two men cross and one returns with his own sister. Then he and I cross over, leaving our sisters behind, and one of the women takes the boat back, and picks up my sister who is carried over to us. Finally, the man whose sister remains on the first bank, crosses over and brings her to us.

This takes a total of eleven crossings, which is more than necessary, as the translator points out. A shorter solution is: I and my sister cross, I return; the other women cross, my sister returns; I and my sister cross again and I return; the other two men cross, and my sister returns; I and my sister cross over. This is a total of nine crossings.

82. 'I would take the goat, and leave the wolf and the cabbage. Then I would return and take the wolf across . . . and take the goat back over; and having left that behind I would take the cabbage across; I would then row again and having picked up the goat take it over once more. By this procedure, there would be some healthy rowing, but without any lacerating catastrophe.'

83. 'First the two children get into the boat, and cross the river; one of them brings the boat back. The mother crosses in the boat, and her child brings the boat back. His brother joins him in the boat and they go across, and again one of them takes the boat back to his father.

The father crosses, and his son ... having boarded, returns to his brother; and both cross again. With such ingenious rowing, the sailing may be completed without shipwreck.'

84. This problem is related to earlier problems (see problem 37 above) based on Islamic and Roman law. The information given is not sufficient, though it might be sufficient if we assume familiar legal principles not stated in the problem.

Alcuin gives the mother the average of the two amounts she would have received in the two cases; the average of $\frac{1}{4}$ and $\frac{5}{12}$ is $\frac{1}{3}$ so the mother receives 320 shillings. The son and the daughter each receive one half of what they would have received if born alone, that is, 360 shillings and 280 shillings respectively.

The translator suggests adding the original fractions they expected, $\frac{3}{4} + \frac{7}{12} + \frac{1}{3}$ (the mother's average expectation) for a total of $\frac{5}{3}$. Now multiplying their original expectations by $\frac{3}{5}$, the mother receives 432 shillings, the boy 336 shillings and the daughter 192 shillings.

85. Alcuin solves this much as Gauss solved, when a small boy in school, the problem of quickly summing the integers from 1 to 100. Alcuin explains: 'Take the one which sits on the first step, and add it to the 99 which are on the 99th step, and this makes 100. Also the second and the 98th, and find again 100. So for each step ... will always give 100 between the two. The fiftieth step is on its own, not having a pair, and similarly the 100th is on its own. Join altogether and get 5050.'

86. There can be no integral solution. However, a rational solution can be found from any three squares in arithmetical progression, whose common difference is of the form $5p^2$, on division by p.

Fibonacci found 31^2, 41^2 and 49^2, whose common difference is $720 = 5 \times 12^2$. Hence his solution is 41/12.

87. Let S be the original sum and $3x$ the sum returned equally to the three men. Before each man received a third of the sum returned, they possessed $\frac{1}{2}S - x$, $\frac{1}{3}S - x$ and $\frac{1}{6}S - x$ respectively. Since these are the sums that they possessed after putting back $\frac{1}{2}$, $\frac{1}{3}$ and $\frac{1}{6}$ of what they had first taken, the amounts first taken were $2(S/2 - x)$, $\frac{3}{2}(S/3 - x)$ and $\frac{6}{5}(S/6 - x)$, and these amounts sum to S.

This gives the equation $7S = 47x$, which is indeterminate, as is inevitable from the original conditions, which only concern proportions with no stated fixed amount.

Fibonacci chose the simplest values, $S = 47$ and $x = 7$. The sums taken by the men from the original pile are then 33, 13 and 1.

[Eves, 1981, pp. 166–7]

88. Assuming that the rabbits are immortal, the number of new pairs produced per month follows this sequence (Leonardo omitted the first term, supposing that the first pair bred immediately):

1   1   2   3   5   8   13   21   34   55   89   144   233   ...

This is the famous Fibonacci sequence, so named by Lucas in 1877. Each term is the sum of the previous two terms. For many of its wealth of properties, see the *Penguin Dictionary of Curious and Interesting Numbers*, p. 61 *et seq*.

Binet proved in 1844 that the nth Fibonacci number is given by the formula:

$$F_n = \frac{(1 + \sqrt{5})^n - (1 - \sqrt{5})^n}{2^n \times \sqrt{5}}$$

89. This is equivalent to a cistern problem. Instead of three pipes pouring water into a pool at different rates, three animals remove flesh from the sheep at different rates.

Fibonacci argues that in sixty hours (a conveniently chosen number) the lion would eat fifteen sheep, the leopard would eat twelve, and the bear ten, a total of thirty-seven. Therefore they will eat one sheep in $60/37 = 1\frac{23}{37}$ hours.

[Fauvel and Gray, 1987]

90. Suppose the last son received N bezants. The last-but-one son, who also received N bezants in total, received $(N - 1) + \frac{1}{7}$ of the remaining bezants at that stage. There were therefore seven bezants remaining after he had taken $N - 1$, so there were $N + 6$ bezants to be distributed after the previous son had received his share, and these $N + 6$ bezants make the two shares of N each taken by the last two sons. Therefore $N = 6$ and, working backwards, the father left an estate of thirty-six bezants, which was divided between six sons.

[Eves, 1976, p. 230]

91. He took 382 apples. The numbers left after he gives one half and one apple more to successive guards, are (382), 190, 94, 46, 22, 10, 4, and 1 for himself.

[Eves, 1976, p. 231]

92. In 63 day-plus-nights it moves upwards $63 \times (\frac{2}{3} - \frac{1}{3}) = 147/5 = 29\frac{2}{5}$.

The final $\frac{3}{5}$ will be covered in $\frac{3}{5} \div \frac{2}{3} = \frac{9}{10}$ of a day, after which the serpent will not slip back. So the total time taken is $63\frac{9}{10}$ days.

93. Draw a circle through the highest and lowest points of the statue, so that it touches a horizontal line through the eye of the spectator. The spectator should stand with his or her eye at that point.

There is some uncertainty in this solution, since the statue is not a vertical rod! Turning the figure on its side, the solution for the rugby conversion problem is the same. The conversion should be taken from T.

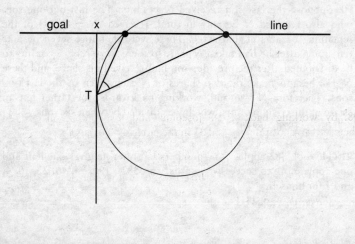

Because of 'the angle in the same segment is equal' property, the same angle will be subtended by the goal-posts at any point on this circle; any point outside it will subtend a smaller angle and any point inside it, a larger angle. (Both true, though not part of the usual statement of the theorem.) Choosing the circle to just touch the line ensures that all other points of the line are *outside* the circle.

94. The speeds of the couriers are 250/7 and 250/9, so their speed of approach to each other is 250 (1/7 + 1/9) = 250 × 16/63 and they will meet in 63/16 = $3\frac{15}{16}$ days.

95. Place two nuns in each of the corner cells, leaving the middle cells empty.

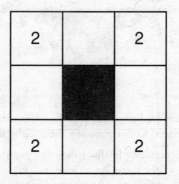

96. He worked for eighteen days and did not work on twelve days.
[Eves, 1976, p. 235]

97. Fill the 5 jar and fill the 3 jar from it, leaving 2 pints in the 5 jar. Empty the 3 jar back into the cask and pour the 2 pints into the 3 jar. Next, fill the 5 jar and fill the 3 jar from it, which takes 1 pint, leaving 4 pints exactly in the 5 jar.

98. By working backwards, Josephus and his companion placed themselves at positions 16 and 31 in the circle of 41 souls.

99. The Christians and Turks should be placed in the following circular order, in which the first person follows the last, and the counting starts with the first person: CCCCTTTTTCCTCCCTCT TCCTTTCTTTCCTT.

100.

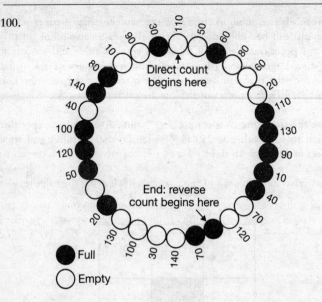

101. As Dudeney pointed out, the central square can be reached by none of the knights, and the other eight squares can be reached from just two other squares, as illustrated by the circular figure.

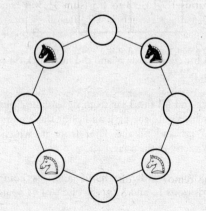

From the circular figure, it is trivially easy to read off the moves that must be made: any piece may make the first move, in either direction, and the other pieces then chug round the circle in the same direction.

102. Tartaglia gives a pheasant to Piero, and to Piero's son Andrea, and to Andrea's son Filippo.

103. Previous authors had assumed that such conditions meant that the inheritance had to be divided literally in the proportions given. It therefore made no difference whether the given fractions added up to less than one, to one exactly, or to more than one.

Tartaglia suggested the device of borrowing an extra horse, making eighteen horses; each person can now be given a whole number of horses, totalling seventeen in all, and the borrowed horse returned to its owner.

A solution by this precise trick method – by borrowing exactly one horse and then returning it – only exists for certain conditions of the problem. If n horses are to be divided among three sons, who receive respectively $1/a$, $1/b$ and $1/c$, then the seven possible values of n,a,b,c are: 7,2,4,8; 11,2,4,6; 11,2,3,12; 17,2,3,9; 19,2,4,5; 23,2,3,8; 41,2,3,7.
[Gardner, 1978a]

However, if you do not mind whether you borrow-and-return or lend-and-recover, and if you do not mind how many horses are borrowed or lent, then the method always works whenever the numerator of the sum of the fractions is equal to the number of horses to be divided. For example, suppose you wish to share 31 horses in the ratios $1/2$ to $1/3$ to $1/5$ (sum 31/30). Lend 1 horse, take $1/2$, $1/3$ and $1/5$ of 30, which is 15, 10 and 6, and then take your borrowed horse back. For $1/2$, $1/3$, $1/4$ and $1/5$, totalling 77/60, share 77 horses by lending 17 and taking them back!

104. Let the barrel originally contain x pints of wine. After one removal and replacement, its strength will be $(x - 3)/x$, and the amount of wine removed on the second removal will actually only be $3(x - 3)/x$, and on the third removal, the wine removed will be

$$\frac{3(x - 3 - 3(x - 3)/x)}{x}$$

The total wine removed is one half the original quantity, and the equation simplifies neatly to $2(x - 3)^3 = x^3$, or $x = (3 \times 2^{1/3})/(2^{1/3} - 1) = 14.54$ pints.

105. If the answer given to you is N, then the number originally chosen is $2N$ or $2N + 1$, depending on whether you were told at the second step that the answer was even or odd, respectively.

106. If the remainders on dividing by 3, 4 and 5, are A, B and C respectively, then the original number is the remainder when $40A + 45B + 36C$ is divided by 60.

Suppose that the number chosen is x, so that the three equations are $x = 3a + A = 4b + B = 4c + C$. Then, multiplying the equations by 40, 45 and 36 respectively:

$$40x = 120a + 40A$$
$$45x = 180b + 45B$$
$$36x = 180c + 36C$$

Adding, $121x =$ a multiple of 60 + $(40A + 45B + 36C)$. Therefore x and $40A + 45B + 36C$ have the same remainder when divided by 60, and since x was chosen to be less than 60, it equals that remainder.

107. Twenty counters. The numbers of counters in the hands of the first and second person are, in three stages: x and $3x$; $x - 5$ and $3x + 5$; $(x - 5) + 3(x - 5)$ and $3x + 5 - 3(x - 5)$, and the last number is always 20.

108. Tartaglia (and Fibonacci before him) had considered the problem of the weights required, if they can only be placed on one side of the balance, and concluded that the best solution has weights in the sequence of powers of 2: 1, 2, 4, 8, 16, 32, and so on. This is the same as saying that each integer can be represented uniquely in the binary notation.

Bachet gave the solution 1, 3, 9 and 27 when both pans may be used. The basic idea is that every number is one more or one less than a multiple of 3. Thus $32 = (3 \times 11) - 1 = 3(3 \times 4 - 1) - 1 = 3(3 \times (3 + 1) - 1) - 1 = 3^3 + 3^2 - 3 - 1$. Therefore 32 lbs can be weighed by placing the 27 and 9 lb weights in one pan and the 3 and 1 lb weights in the other.

It is plausible that Bachet's solution is in some sense best possible, merely because it is so simple and elegant. This was proved in 1886 by Major MacMahon, who used the method of generating functions discovered by Euler to show that there are eight possible sets of weights, apart from the one-scale solution, $1, 2, 4 \ldots 32$. Denoting the number of each weight by a superscript, they are:

1^{40}; $1, 3^{13}$; $1^4, 9^4$; $1, 3, 9^4$; $1^{13}, 27$; $1, 3^4, 27$; $1^4, 9, 27$; $1, 3, 9, 27$.

Thus Bachet's solution does indeed use the fewest weights and is also the only solution in which all the weights are different.

109. The easiest way to construct Heronian triangles is to fit two right-angled triangles together.

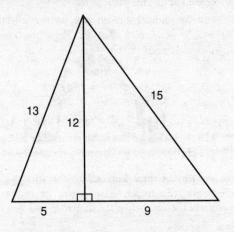

Here are a 5–12–13 and 9–12–15 triangle fitted together. The resulting triangle has altitude and sides 12–13–14–15, and is the only possible such triangle. The area is 84.

110. The right-angled triangles 5–12–13 and 6–8–10 each have area equal to perimeter. The three proper Heronian triangles with this property are 6–25–29; 7–15–20; 9–10–17.

111. In the solution to 109, place the same two triangles so that they overlap. The obtuse-angled triangle with sides 4–13–15 has area 54 − 30 = 24.

Proofs of these results are not so simple; one method is to write Heron's formula in the form

$$16A^2 = 2a^2 b^2 + 2b^2 c^2 + 2c^2 a^2 - a^4 - b^4 - c^4$$

which can be written in the form

$$(4A)^2 + (b^2 + c^2 - a^2)^2 = (2bc)^2$$

This is of the form $p^2 + q^2 = r^2$ and has parametric solutions, $p = m^2 - n^2$, $q = 2mn$ and $r = m^2 + n^2$.

112. Each knife blade goes under the blade of one other knife and over the blade of the third. So arranged, they can easily support a glass of water well above the table surface.

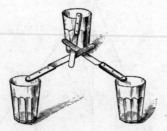

113. Force the tips of three knives into the stick, so that the knives hang well below the finger. The centre of gravity of the entire arrangement will then be below the finger tip and will be stable.

114.

115. Bend the straw and insert into the neck of the bottle, which can then be lifted.

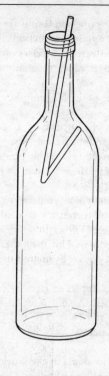

116.

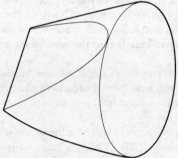

Van Etten gives variants of this puzzle, in which the combination of holes to be plugged are different, such as square, circular and oval. The principle is the same.

117. He is standing at the centre of the earth.

118. Van Etten gives the same solution – they are at the centre of the earth, and ascending in opposite directions. However, it is also true that if they ascended two vertical ladders on the earth's surface, they would also be moving apart, albeit by a minuscule amount.

119. When he is standing at the North Pole.

120. Place the point on the surface of a sphere and draw a circle, which will be smaller than the circle drawn by the same compass on a plane surface. Alternatively, place the point of the compass at the apex of a circular cone, and draw a circle on its surface.

121. Wrap the paper on which you are to draw the oval round a cylinder. The compass will then draw an oval.

122. One horse travelled east and the other travelled west, the first gaining in the number of days it lived, and the second losing.

123. First they sold their apples at 1 penny each, then later in the day they sold them at 3p each. A sold 2 @ 1p and 18 @ 3p, making 56 pence. B sold 17 @ 1 p and 13 @ 3 p, and C sold 32 @ 1 p and 8 @ 3p, each also making 56 pence.

124. The number of individuals in the world far exceeds the number of hairs on the head of any one of them. Therefore if you start to pick out individuals with given numbers of hairs on their heads, you will be forced to pick an individual with a number of hairs that you have already counted once, long before the population of the world is exhausted.

This is the first known example of the 'pigeonhole principle', which says that if you have $N + 1$ objects to place in N pigeon holes, then one hole must contain at least two objects.

125. The sum of the distances will be a minimum when the lines OA, OB and OC all meet at 120°. (This is a general principle that applies to all such minimum networks of 'roads'. If one of the angles is greater than or equal to 120°, the point sought is at that vertex.)

To construct point O with ruler and compasses, draw equilateral triangles outwards on each side of the triangle, and draw the circumcircles of each new triangle. These circles will pass through a common point, which is O. This construction works, because of the property that the opposite angles of a cyclic quadrilateral sum to 180°. Choos-

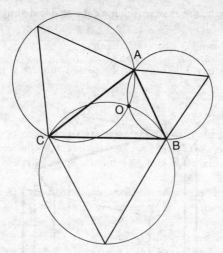

ing the extra triangles to be equilateral, with angles of 60°, ensures that the opposite angles will all be 120° as desired.

It also happens that if the outer vertices of the equilateral triangle are joined to the opposite vertices of the original triangle, they will all pass through the point O.

[For further discussion see Honsberger, 1973, p. 24]

126. It is only necessary to 'tie' an ordinary knot in the strip and carefully flatten it.

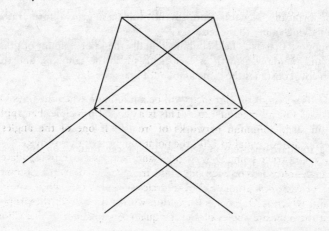

127. Amazingly, the answer is that a cube of side slightly under

$$\frac{3\sqrt{2}}{4}$$

or approximately 1.060660, can be passed through a given cube of side 1 unit.

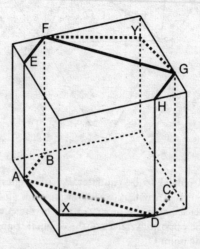

In this figure, the square hole cuts the top face along the lines EFGH, the bottom face along ABCD and the two vertical edges at X and Y, as indicated by the dotted lines.

128. Typically, Newton expresses the problem in general form, rather than giving values to the letters.

Suppose that each field contains initially the same amount of grass, M, and the daily growth in each field is also the same, m, and that each cow consumes the same amount of grass per day, Q.

Then
$$bM + cbm - caQ = 0$$
and
$$b'M + c'b'm - c'a'Q = 0$$
and
$$b''M + c''b''m - c''a''Q = 0$$

By a standard theorem in determinants, given that M, m and q are not all zero, this determinant is zero:

$$\begin{vmatrix} b & bc & ca \\ b' & b'c' & c'a' \\ b'' & b''c'' & c''a'' \end{vmatrix}$$

Without using determinants, M, m and Q can be eliminated 'by hand', to give

$$b''cc'(ab' - ba') + c''b''(bc'a' - b'ca) + c''a''bb'(c - c') = 0$$

[After Dorrie, 1965, p. 9]

129. It might seem that the chances are equal, because the proportion of sixes required to the number of dice thrown is constant. This is not so.

'The chance of getting 1 six and 5 other outcomes in a particular order is $(\frac{1}{6})(\frac{5}{6})^5$. We need to multiply by the number of orders for 1 six and 5 non-sixes. Therefore the probability of exactly 1 six is

$$\binom{6}{1}\left(\frac{1}{6}\right)\left(\frac{5}{6}\right)^5$$

Similarly, the probability of exactly x sixes when 6 dice are thrown is

$$\binom{6}{x}\left(\frac{1}{6}\right)^x\left(\frac{5}{6}\right)^{6-x}, \qquad x = 0, 1, 2, 3, 4, 5, 6$$

The probability of x sixes for n dice is

$$\binom{n}{x}\left(\frac{1}{6}\right)^x\left(\frac{5}{6}\right)^{n-x}, \qquad x = 0, 1, \ldots, n$$

This formula gives the terms of what is called a binomial distribution.

'The probability of 1 or more sixes with 6 dice is the complement of the probability of 0 sixes:

$$1 - \binom{6}{0}\left(\frac{1}{6}\right)^0\left(\frac{5}{6}\right)^6 \approx 0.665$$

'When $6n$ dice are rolled, the probability of n or more sixes is

$$\sum_{x=n}^{6n} \binom{6n}{x}\left(\frac{1}{6}\right)^x\left(\frac{5}{6}\right)^{6n-x} = 1 - \sum_{x=0}^{n-1} \binom{6n}{x}\left(\frac{1}{6}\right)^x\left(\frac{5}{6}\right)^{6n-x}$$

Unfortunately, Newton had to work the probabilities out by hand, but we can use the *Tables of the Cumulative Binomial Distribution*, Harvard University Press, 1955. Fortunately, this table gives the cumulative binomial for various values of p (the probability of success on a single trial), and one of the tabled values is $p = \frac{1}{6}$. Our short table shows the probabilities, rounded to three decimals, of obtaining the mean number or more sixes when 6n dice are tossed.

6n	n	P(n or more sixes)
6	1	0.665
12	2	0.619
18	3	0.597
24	4	0.584
30	5	0.576
96	6	0.542
600	100	0.517
900	150	0.514

Clearly Pepys will do better with the 6-dice wager than with 12 or 18. When he found that out, he decided to welch on his original bet.'
    [Mosteller, 1987, problem 19]

130. Bernoulli posed the problem in terms on n letters wrongly placed into n envelopes, but the principle is the same. The general formula for the number or ways of misplacing all the letters is:

$$n!\left(\frac{1}{2!} - \frac{1}{3!} + \frac{1}{4!} - \frac{1}{5!} + \ldots + \frac{-1^n}{n!} \right)$$

When $n = 7$, the value is 1854.
    [A complete solution is in Dorrie, 1965, p. 19]

131. One! This answer is independent of the number of letters and envelopes.
    [Newman, 1982, p. 22]

132. These two solutions have the features of simplicity and symmetry.

[Fisher, 1973]

133. No, it is not. The map of the river can be represented schematically like this:

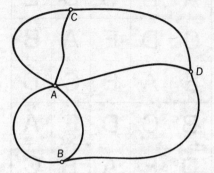

The problem is now to trace out this figure with a pencil, passing over every line once, and no line twice, without lifting your pencil from the paper. This is only possible if the figure to be traced contains either no vertices at which an odd number of edges meet (in which case you may start at any point you choose, and trace the figure so as to return to your starting point), or just two such vertices, in which case you can only trace the figure by starting at one and ending at the other.

The reasoning is simple: in arriving at a vertex and then leaving it, two of the edges meeting at the vertex are 'used up'. Therefore, any vertex at which an odd number of edges meet (all of which must be traversed) can only be a starting or an ending vertex, of which there can be at most two.

All four vertices in the figure for the Bridges of Königsberg are 'odd', and so the figure cannot be traversed.

134. Every edge has two ends, so the total number of edge-ends is even. But the number of edge-ends is also the total of the number of edges meeting at each of the individual vertices, which must therefore include an even number of odd vertices, since an odd number of odd vertices would give an odd grand total.

This was one of the points established by Euler in his original paper.

135. This is one solution. Typically the same letters are knight's moves apart from each other, and the patterns formed by the shading are similar.

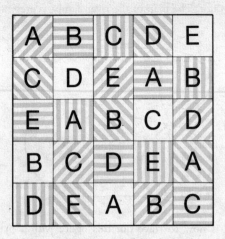

136. 19,013 years, 144 days, 5 hours and 55 minutes.

137. Sixty-six years.

138. Forty-five and fifteen years.

139. The given solution is something of a cheat, quickly referring to one of Dr Hutton's textbooks. (Readers also occasionally stooped to taking questions from published sources. As the editor remarks of Question 97, 'It is evident that this question is composed from that in page 225 of Ward's *Mathematician's Guide*'!)

Assuming that the cylinder to be cut out has its base on the base of the cone, it remains only to determine its height, as a proportion.

The volume of the cylinder is proportional to $DE^2 \times PQ$, and we know that DE/AP is constant. Therefore, it is required to maximize $AP^2 \times PQ$. In other words, given any line AQ, find a point on it, P, such that $AP^2 \times PQ$ is a maximum.

For clarity, call AP x, and let the height $AQ = L$. Then we have to maximize $x^2(L - x)$, which is the same as maximizing $x^2(2L - 2x)$. But the latter is the product of three factors whose sum is constant, at $2L$, and it will achieve its maximum value if all three factors are equal. Therefore the maximum is when $x = 2L - 2x$ and $x = 2L/3$. So AP must be two-thirds of AQ.

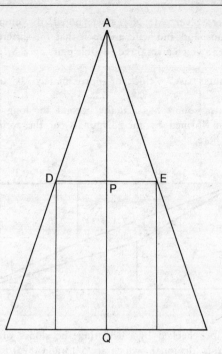

140. 2520 = 5 × 7 × 8 × 9.

141. This is the solution by Mr J. Hill: 'Call the number of hogs any [one] woman bought x; the number her husband bought $x + n$; money laid out by the woman is xx shillings; money laid out by the husband is $xx + 2nx + nn$ shillings. Equation

$$xx + 2nx + nn = xx + 63$$

[Therefore] $x = (63 - nn)/2n$.

If $n = 1$, then $x = 31$ and $x + n = 32$; hence some woman bought 31 hogs, and her husband 32. If $n = 3$, then $x = 9$, and $x + n = 12$; therefore some other woman bought 9, and her husband 12. If $n = 7$, then $n + x = 8$; [therefore] some woman bought 1, and her husband 8. Consequently,

Hendrick bought 32 and his wife Anna 31
Claas . . . . 12 . . . . Catriin 9
Cornelius . . . . 8 . . . . Geertrick 1'

As another solver, Mr N. Farrer, noted, the numbers of hogs bought by husband and wife are such that the differences of their squares are 63, which gives three possible pairs only: 8,1; 12,9; 32,31.

142. $5\frac{5}{11}$ minutes past 7 o'clock, or approximately 7.05 and 27 seconds.

143. The extra square has actually become the long and very thin parallelogram formed by the 'diagonals' of the second figure, as exaggerated here.

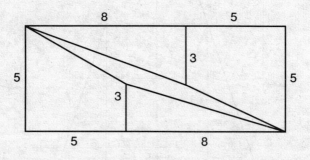

This can be checked by calculating the slopes of the different portions of this 'diagonal', which are 5/13 and (8–5)/8 = 3/8. These fractions are close, but not equal.

The fractions are so close because they have been conveniently taken from the Fibonacci sequence, 1 1 2 3 5 8 13 21 34 55 ... which has the property that of any four consecutive terms, the products of the outer pair and of the inner pair differ by one. By taking four consecutive terms later in the sequence, the paradox becomes even more difficult to detect by eyesight alone.

144.

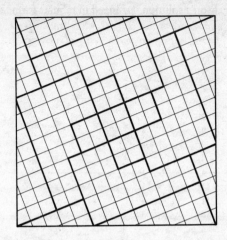

[Delft and Bottermans, 1978]

145. The rowers move at 4 miles per hour relative to the water, so they approach each other at 8 miles per hour, and will close the 18 miles between them in 2¼ hours. If the water were still, their meeting would be at the mid-point, 9 miles from each town. But due to the current, their meeting place will have moved at 1½ miles per hour, over the 2¼ hours, downstream from Haverhill towards Newburyport, a total drift of 3⅜ miles. So they meet 5⅝ miles from Newburyport.

146. 99⅔.

147.

	SIX		IX		S
From	IX	take	X	leaving	I
	XL		L		X

148. 2 + 4 + 6 + 0.8 = 12.8 and 1 + 3 + 7 + 9/5 = 12.8.

149. One-third of TWELVE is LV = 55 in Roman numerals; one-fifth of SEVEN is V = 5, and 55/5 = 11.

150. 123456789 × 8 = 987654312.

151. This is Jackson's solution, arranged in tabular form:

barrel	first container	second container
12		
7	5	–
7	–	5
2	5	5
2	3	7
9	3	–
9	–	3
4	5	3
4	1	7
11	1	–
11	–	1
6	5	1
6	–	6

152. If cells are filled with the numbers (reading left to right, top to bottom) 7, 0, 5; 2, 4, 6; 3, 8, 1, then the square is magic in the usual way, by addition of the rows and columns. If each number is replaced by the matching power of 2, then it is magic by multiplication of rows and columns. So, one solution, reading the rows from left to right, top to bottom, is: 128, 1, 32; 4, 16, 64; 8, 256, 2.

153. This is Jackson's solution:

4	9	5	16
15	6	10	3
14	7	11	2
1	12	8	13

154. Jackson merely states that the true weight is the mean proportional (or geometric mean), that is, the square root of 16 × 9 or 12 lbs.

Suppose that the long and short arms of the balance are of length p and q respectively, and the true weight is W. Then $Wp = 16q$ and $Wq = 9p$, from which $W^2pq = (16 \times 9)pq$ and the conclusion follows.

155. A shoe.

156. This is Jackson's solution. It generalizes to dividing a circle into N parts by N lines of equal length. Divide a diameter into N equal parts, and construct a sequence of semi-circles on either side, following this pattern. The tadpole-shaped regions at each end and the $N - 2$ snake-like regions between will all have equal area, and will be bounded pairs of lines each equal in length to half the perimeter of the circle.

157. Draw it on a sphere, taking, for example, one of the poles and any two points on the equator which are separated by one quarter of the earth's circumference (taking the earth to be spherical).

158. O is the centre of the circle. With the compass open to the radius of the circle, mark off the points C, X, and B in succession. Then with radius BC, and centres A and B, draw arcs to intersect at D.

Then DO is the length of the side of the required square. Marking AE so that AE = DO, and similarly marking F, produces the square.

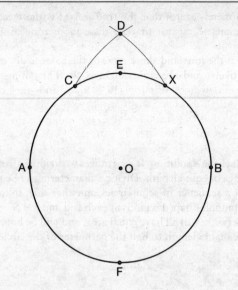

Also, since AOE = 90° and AOC = 60°, COE = 30° and CE is one side of an inscribed dodecagon.

159. 'Suppose one place to lie directly under either of the poles, a second 10 degrees on this side, and a third 20 degrees on the other, under the same meridian circle, then they will all differ in latitude, and likewise in longitude, since the pole contains all degrees of longitude.'

The figures chosen by Jackson are, of course, quite arbitrary.

160. The South Pole.

161. The reference to Naples and the situation of the village in a low valley are mere flim-flam, worthy of Sam Loyd. The fact is that any place on earth, the poles apart, varies daily in distance from the sun because of the earth's rotation, being a maximum for places on the equator, and a minimum of zero for the actual poles. 3000 miles is Jackson's estimate of the variation for Naples, based on the earth having a radius of about 4000 miles.

162. The island is Guernsey. (Any of the other Channel Islands would do.) Guernsey is 26 miles from France, but England, from Dover to Calais, is only 21 miles from France.

163. The Christian sets off from the Jew's abode, travelling East, and the Turk does likewise, but travels West. When they meet again at the Jew's, by their own reckoning, having respectively lost a day and gained a day while travelling round the world, they will each be able to celebrate their own sabbath on the same day in the same place!

164. The traveller's journey has been right round the world. His head is about 6 feet from the ground, and so the radius of the giant circle travelled by his head is about 6 feet greater than that of the circle travelled by his feet. This difference in radius produces a difference in circumferences of about $2\pi \times 6$ feet, or about 36 feet = 12 yards.

   Jackson attributes this idea to Whiston's commentary to his edition of Euclid.

165. At $1\frac{1}{11}$, $2\frac{2}{11}$, $3\frac{3}{11}$ ... hours.

166. There are fourteen different arrangements, ignoring colours, each of which can be coloured in two ways, making a total of twenty-eight. (Some pairs of these arrangements would be equivalent if the tiles were painted on both sides, and arrangements of tiles could be turned over.)

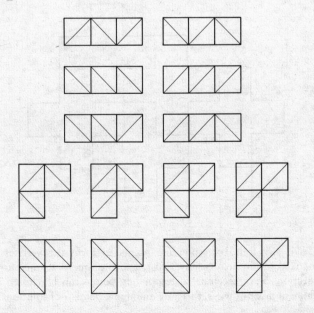

167. A: $14\frac{34}{49}$ B: $17\frac{23}{41}$ C: $23\frac{7}{31}$

168. This solution is equivalent to the problem of dissecting a Greek Cross into a square: assemble the five squares to form the cross, and the solution appears immediately.

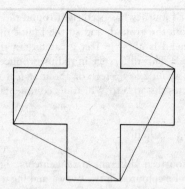

This dissection of a Greek Cross is one of an infinite number which depend on the possibility of tessellating the plane with identical crosses:

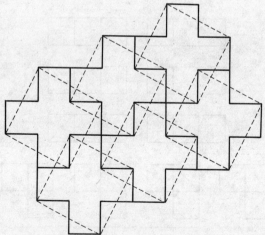

Take any four corresponding points to form the vertices of a square, equal in area to any Cross, and the square at once forms a dissection. Since the four corresponding points can be chosen in an infinite number of ways, there are an infinite number of solutions.

169. The two smaller squares in the Pythagoras figure will tessellate also, and by joining corresponding points together, an infinite number of dissections of the smaller squares into the larger are found. In every case, the pieces require only to be slid, without rotation, into their new positions.

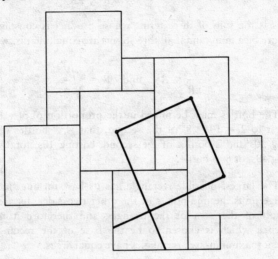

170.

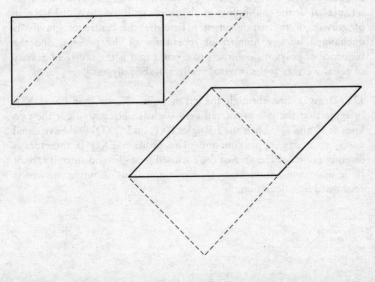

171. Let the radius of the base be R and the height H. Then the total surface area is $\pi R^2 + 2\pi RH$, and the volume, which is fixed, is $V = \pi R^2 H$.

Therefore the surface area is $\pi\left(R^2 + \dfrac{V}{\pi R} + \dfrac{V}{\pi R}\right)$.

This is the sum of three terms whose product is constant. It will therefore be a minimum if all three terms are equal, that is, when

$$\frac{V}{\pi R} = R^2, \quad \text{and} \quad R = \sqrt[3]{\frac{V}{\pi}}.$$

172. The bottles must be mixed in the proportion of $3(= 8 - 5)$ of the first to $2(= 10 - 8)$ of the second, that is, 3 bottles of the first costing 30s and 2 bottles of the second, costing 10s; total, 5 bottles costing 40s, or 8s a bottle.

173. The largest possible rectangle has its base on one side, and its top edge joins the mid-points of the other two sides. Its area is then one half of the area of the triangle, and therefore it makes no difference which is chosen to be the base of the rectangle: three different maximum-area rectangles have equal areas.

174. Imagine that equal weights are placed at the vertices of the original polygon. Then replacing these equal weights by an identical set, placed at the mid-points of the sides, will not change the centre of gravity of the arrangement. Therefore the centre of gravity is unchanged by any number of repetitions of the process, and the sequence of polygons contracts to a point that is the centre of gravity of equal weights at the original vertices of the polygon.

175. Draw a line through the given point, X, so that PX = XQ. Imagine that the line is rotated very slightly about X to cut the two lines at P' and Q'. Then the triangles PXP' and QXQ' will have equal areas, to a first approximation. The position PXQ is therefore a limiting point for the area of the enclosed triangle, and since it cannot be a maximum, it must be a minimum. This intuitive answer is confirmed by calculation.

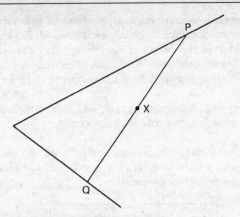

176. Ozanam gives the solution square in this algebraic form, so that any numbers can be substituted for *a*, *b* and *c*.

a	$\dfrac{2ac}{a+c}$	c
$\dfrac{2ab}{a+b}$	$\dfrac{2bc}{b+c}$	$\dfrac{2abc}{2ab+ac-bc}$
b	$\dfrac{2abc}{2ac+ab-bc}$	$\dfrac{abc}{ab+ac-bc}$

To get whole-number solutions, he chooses values of *a*, *b* and *c* to give this square:

1260	840	630
504	420	360
315	280	252

177. The secret is to make the sum, after one of your turns, equal to a number in the sequence 1, 12, 23, 34, 45, 56, 67, 78, 89, 100. Once you have achieved this (which is easy enough if you start the game – you just choose 1 as your first call), then you can keep to the sequence by calling out the difference between 11 and your opponent's last call.

If your opponent starts the game, and knows the trick, then of course you must lose, but only alternate games!

178. Every solution can be rotated and reflected to reproduce seven other solutions, which, however, will in some cases be identical to the original solution due to its symmetries.

There are twelve basic solutions on the full chessboard. Each can be described by a single 8-digit number, by reading off the position of each queen in each column, starting from one end. With this notation, the twelve solutions are: 41582736; 41586372; 42586137; 42736815; 42736851; 42751863; 42857136; 42861357; 46152837; 46827135; 47526138; 48157263. [Rouse Ball, 1974, p. 171]

179. Using the same notation as solution 178, there is one basic solution on a 4 × 4 board: 3142. There are two basic solutions on a 5 × 5: 14253 and 25314; and one on a 6 × 6: 246135.

On a 7 × 7 board there are six basic solutions.

180. This is the commonest solution offered:

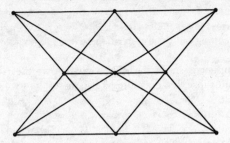

It is a special case of Pappus's theorem, which says that if points A, B and C are taken on one line, and A′, B′ and C′ taken on another, and joined as in the figure, then the points P, Q and R will lie on a line. (P is the meet of BC′ and B′C, and so on.) If it happens that B, Q and B′ also lie on a line, then the figure satisfies the conditions of the problem.

A more general solution is any figure for Desargues's theorem. In other words, take any two triangles, such that the lines through matching vertices meet in a point P.

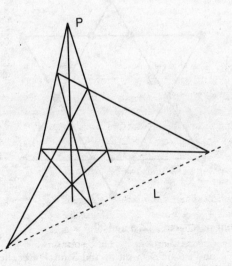

The pairs of corresponding sides will then also meet in pairs on a line L.

181.

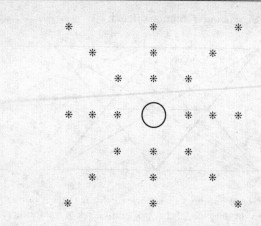

182. Plant three of the trees at the base of a steep mound, at equal distance from each other, and plant the fourth tree on top, at equal distance from the other three.

183. This solution (ignoring the dotted lines) can be varied, by varying the triangles. Indeed, any two overlapping triangles will do.

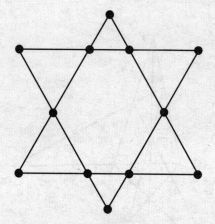

184. The four parts are 8, 12, 5 and 20.

185. 'Place 4 on 7, 6 on 2, 5 on 8, and 3 on 1. Recollect always to begin with either 4 or 5. The same trick may be thus performed: place 5 on 2, 3 on 7, 8 on 6, and 4 on 1.'

186.

187.

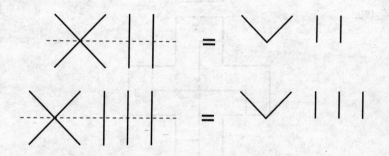

188. This puzzle was the basis for one later made famous by Sam Loyd.

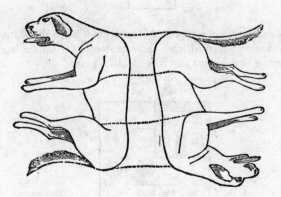

189.

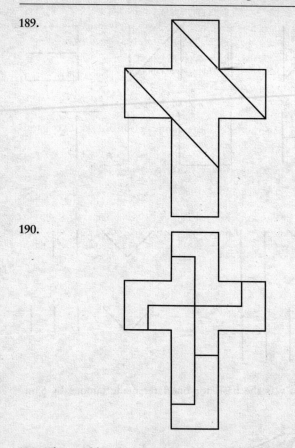

190.

191. This problem has reappeared many times, often in its simplest guise – an L-shape formed from three-quarters of a square, and the demand that it be dissected into four identical parts.

192. Fold one edge of the square on to its opposite edge to get a middle line. With the middle line horizontal, fold one lower corner on to the middle line, so that the fold passes through the other lower corner. Repeat, using the other lower corner. These two folds and one edge form an equilateral triangle.

['Tom Tit', n.d.]

193. 'Roll the paper into a short compact roll. Make two parallel cuts across the roll, each being about one half an inch from the other end. Then make a long cut parallel to the axis of the roll and terminated by the cross cuts. This will produce a gap in the roll. Holding the roll lightly in the fingers ease out the ends of the first strip, which lies at the bottom of the gap, then, taking the strip with the teeth and holding the roll lightly by its two ends, slowly draw the strip out of the gap ... the whole inside of the roll will be drawn through the gap, the connecting parts of the successive strips being twisted. The final result will be a series of paper strips which serve as rungs of a ladder, whose upright sides are formed by the twisted parts.'

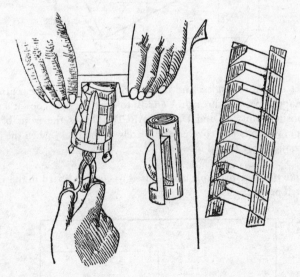

194. Trace round a shilling and cut out the circle to make a circular hole. This hole cannot be increased in size as long as the paper is flat, without tearing the paper. However, if the paper is folded across a diameter of the circular hole, and a larger coin – the half-crown was just the right size – is placed within the fold next to the hole, then by bending the paper without tearing, the hole can be enlarged sufficiently to allow the larger coin through. The maximum diameter of the hole when the paper is bent is $\sqrt{2}$, or very slightly over 1.4, times its original diameter.

195. Draw a rectangle inside the given sheet so that there is a sufficient margin for gluing the resulting envelope.

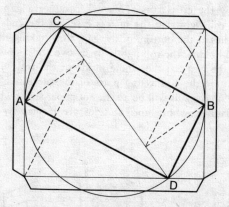

Bisect the shorter sides, and draw a circle, centre at the centre of the rectangle, passing through A and B, to cut the other sides at a pair of opposite points, C and D. Join ABCD. This is the front of the envelope, which is just covered completely at the back, when the four outer triangles are folded inwards.

196. Like this (the principle can be used to cut any board in the same or related proportions):

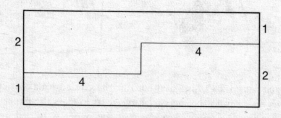

197.

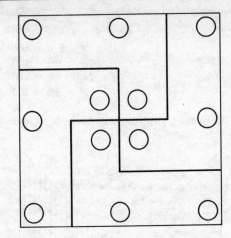

198. Fold the card in half, and make cuts with scissors as shown. Finally cut down the original central fold, omitting the two end portions. The card is now reduced to a strip which may be opened out and passed over a person.

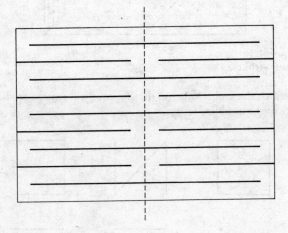

199. Number the coins from 1 to 10, in sequence. Place 4 on 1, 7 on 3, 5 on 9, 2 on 6, and then 8 on 10.

200. Here are a handful of solutions:

$$123 - 4 - 5 - 6 - 7 + 8 - 9 + 0 = 100$$
$$1\tfrac{3}{6} + 98\tfrac{27}{54} + 0 = 100$$
$$80\tfrac{27}{54} + 19\tfrac{3}{6} = 100$$
$$70 + 24\tfrac{9}{18} + 5\tfrac{3}{6} = 100$$
$$87 + 9\tfrac{4}{5} + 3\tfrac{12}{60} = 100$$

201. Giraffe, lion, camel, elephant, hog, horse, bear, hound.

202.

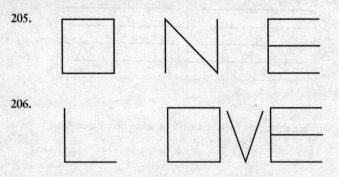

203. Four strokes.

204. Scratch the table cloth and the 20p coin will emerge from under the edge of the tumbler.

205.

206.

207. Before picking up the handkerchief, fold your arms. Then pick up the handkerchief, and unfold your arms. The 'knot' which was in the folding of your arms is transferred, as it were, to the handkerchief.

208. Fold the piece of paper on which the sum is written, like this, so as to obscure the figure 300, and so that the other two lines form the new number 707. The new sum has the same total as the old.

$$
\begin{array}{r}
318 \\
\cancel{707} \\
215 \\
\hline
1240
\end{array}
$$

209. Jump 9 over 4, 5, 7, and 1; 3 over 2; 6 over 8 and 3; and 9 over 6.

210. Arrange the matches to form the edges of a regular tetrahedron.

211.

212.

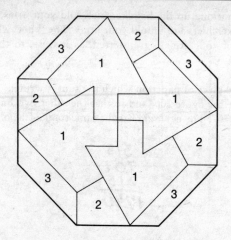

213. Jane, Ann, Joe, Bet, Rose and Jim earn, respectively, 3s 2d, 2s 7d, 1s 11d, 1s 5d, 1s 1d and 8d per week.

214. The number of letters contained in each numeral word.

215. The four figures are 8888, which on being divided horizontally along the middle line become a row of zeros, or nothing.

216. (a) $19\frac{1}{2}$

$$191/2$$

    (b) Place one of the coins on the table, then keeping the hands apart, take it up with the other hand.

    (c) Draw it round his body.

    (d) $8\frac{1}{4}$.

217. The squirrel takes out each day one ear of corn and his own two ears.

218.

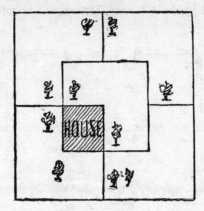

219. The difference in their ages.

220.

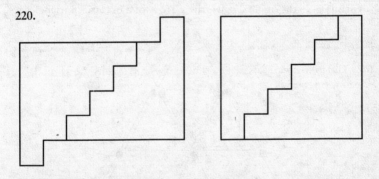

221. $35/70 + 148/296 = \frac{1}{2} + \frac{1}{2} = 1$

222. The original puzzle states that the two digits transferred are 28. The answer is 285714, which is the period of the decimal fraction $\frac{2}{7} = 0.\dot{2}8571\dot{4}$.

This period has the property that any circular rearrangement of the digits is a multiple of the period of $\frac{1}{7}$, 142857.

223. 'The reply of most people is, almost invariably, that the hatter lost £3 19s 0d and the value of the hat, but a little consideration will show that this is incorrect. His actual loss was £3 19s 0d less his trade profit on the hat; the nett value of the hat, plus such trade profit, being balanced by the difference, 21s, which he retained out of the proceeds of the note.'

224.

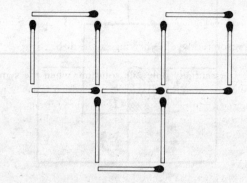

225. Remove the dotted matches, and three small triangles, three medium-sized and the outer triangle are left, a total of seven triangles remaining. In the original figure there are a total of thirteen triangles.

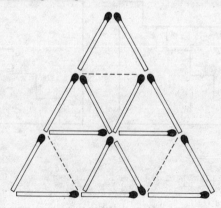

226. The old gentleman was a widower with a daughter and sister. The old gentleman and his father (who was also a widower) married two sisters (the wife of the old gentleman having a daughter by a former husband); the old gentleman thus became his father's brother-in-law. The old gentleman's brother married the old gentleman's step-daughter; thus the old gentleman became his brother's father-in-law. The old gentleman's father-in-law married the old gentleman's sister, and the old gentleman thus became his father-in-law's brother-in-law. The old gentleman's brother-in-law married the old gentleman's daughter, whereby the old gentleman became his brother-in-law's father-in-law.

227. $864 - 72 = 792$.

228. One person received his herring on the dish.

229. There is essentially only one solution, when the symmetries of the dodecahedron are taken into account, which is easily read off the plane map. For example, go round the minor pentagon, move to the 'star' pentagon and go round it, and then end by going round the outer pentagon.

230. However the dogs run, the distance between each dog and the dog he is chasing will be reduced from the initial 100 yards at the rate of 3 yards per second. They will therefore meet in $33\frac{1}{3}$ seconds, and by symmetry they will collide at the centre of the field.

231. De Morgan was born in 1806, and so was 43 in the year 1849 = 43^2.

232. This is a minimum solution, in forty-six moves, due to Dudeney. Note that it is generally only necessary to name the piece that moves at each turn. The '*' symbol indicates that a piece has jumped, leaving the central cell empty:

Hhg*Ffc*CBHh*GDFfehbag*GABHEFfdg*Hhbc*CFf*GHh*

233. The maximum value which cannot be made is 63. Higher numbers can always be made; thus, $64 = 2 @ 17 + 6 @ 5$.

234. Yes. Move B one square to the right and move A round the circuit to the right of B. Interchange C and D by moving round the shaded cells and then shunt B–A to the left.

235. First, train B advances, and backs its rear half into the cul-de-sac, uncouples it, and moves well forward of the junction. Second, A passes the junction, backs, and joins to the rear half of B, which it then draws out of the cul-de-sac and backs to the left. Third, the front half of train B backs into the cul-de-sac. Fourth, train A uncouples the rear half of train B and proceeds on its way. Train B can now leave the cul-de-sac and join its rear half and proceed.

236. No. The planks can be used together, provided that their length is at least

$$\frac{2\sqrt{2}W}{3},$$

where W is the width of the moat (with a little allowance for the planks to overlap each other and the bank).

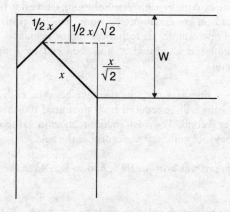

Since $8 < \dfrac{2\sqrt{2} \times 10}{3}$ the given planks will not suffice.

237. The solution can be represented visually, like this:

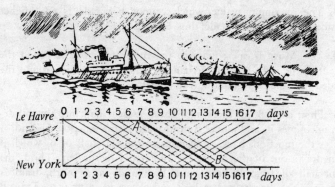

The line AB represents the ship leaving Le Havre today. It passes thirteen ships at sea, and meets two more ships, one in each harbour, a total of fifteen.

[Attributed to Lucas in Kordemsky, 1972, problem 255]

238. The principles are the same whatever the number of discs. Suppose, therefore, for simplicity, that there are eight discs to be moved, from peg A to peg B. Number the discs 1 to 8 from the top downwards. The simplest rules are:

1. Always move an odd numbered disc, *on its first move from peg A*, to peg C, and an even numbered disc, *on its first move from peg A*, to peg B.
2. Move disc 1 every second move, disc 2 every fourth move, disc 3 every eighth move, and so on.

Following this rule, the sequence for eight discs will start: 1–C, 2–B, 1–B (the first two discs have now been transferred to peg B, solving the problem for just two discs); 3–C, 1–A, 2–C, 1–C (leaving discs 1 to 3 on C); 4–B, 1–B, 2–A, 1–A, 3–B, 1–C, 2–B, 1–B (leaving the first four discs on peg B).

Notice that disc 1 visits the pegs repeatedly in the order C–B–A–C–B–A . . .; disc 2 visits them in the order B–C–A–B–C–A so that its visits 'rotate' in the opposite direction, and similarly for the remaining discs.

The next move is to place disc 5 on to peg C and repeat the process so far, to leave all discs up to 5 on C. Then place disc 6 on B and repeat to get all discs up to 6 on B, place disc 7 on C and repeat to get all discs up to 7 on C, and finally put disc 8 on peg B and repeat the entire process to transfer the seven smaller discs from peg C to peg B.

The number of moves taken to move $n + 1$ discs is one more than twice the number needed to move n discs. It is therefore $2^n - 1$. In Lucas's original story the number of moves required is therefore $2^{64} - 1$, which at one move every second amounts to more than 500,000,000,000 years.

239. The probability is $\frac{2}{3}$. The following explanation is due to Howard Ellis, one of Gardner's readers. Let B and W^1 stand for the black or white counter which is in the bag at the start and W^2 for the added white counter. After removing a white counter there are three equally likely states:

In bag	Outside bag
W^1	W^2
W^2	W^1
B	W^2

In two out of three cases, a white counter remains in the bag.
[Carroll, 1958, problem 5, and Gardner, 1981, p. 189]

240. This is equivalent to asking for the relative volumes of a regular tetrahedron and a regular octahedron. Carroll solved this problem by calculation, but it is solved more efficiently by visualization. Fit two such pyramids together to make a regular octahedron and inscribe it in a regular tetrahedron.

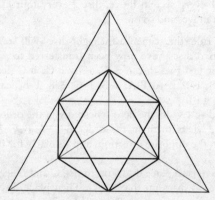

The complete tetrahedron has $2 \times 2 \times 2 = 8$ times the volume of any of the four small tetrahedra affixed to alternate faces of the octahedron. Those four tetrahedra therefore occupy in total one half of the volume of the large tetrahedron, and the octahedron occupies the other half.

The pyramid which is one half of the octahedron therefore occupies one quarter of the complete tetrahedron and is equal in volume to two of the small tetrahedra.

[Carroll, 1958, problem 49]

241. This is Carroll's own solution:
'It may be assumed that the three points form a triangle, the chance of their lying in a straight line being (practically) nil.

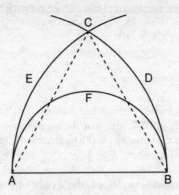

'Take the longest side of the triangle and call it *AB*; and on that side of it on which the triangle lies draw the semicircle *AFB*. Also, with centres *A*, *B*, and distances *AB*, *BA*, draw the arcs *BDC*, *AEC*, intersecting at *C*.

'Then it is evident that the vertex of the triangle cannot fall outside the Figure *ABDCE*.

'Also, if it falls inside the semicircle, the triangle is obtuse-angled; if outside it, acute-angled. (The chance of its falling *on* the semicircle is practically nil.)

'Hence required chance $= \dfrac{\text{area of semicircle}}{\text{area of Fig. } ABDCE}$.

'Now let $AB = 2a$: then area of semicircle $= \dfrac{\pi a^2}{2}$; and area of Figure $ABDCE = 2 \times$ sector $ABDC -$ triangle ABC;

$$= 2\left(\frac{4\pi a^2}{6}\right) - \sqrt{3}a^2 = a^2\left(\frac{4\pi}{3} - \sqrt{3}\right);$$

$$\therefore \text{ chance} = \frac{\pi/2}{\frac{4}{3}\pi - \sqrt{3}} = \frac{3}{8 - \frac{6\sqrt{3}}{\pi}}.$$

[Carroll, 1958, problem 58]

242. The clock that is losing time is correct once every two years, whereas the stopped clock is right twice a day, every time that 'its' time comes round!

[Carroll, 1961, p. 6]

243. Ten. Adding the four percentages together, the total is 310 per cent. Distributing them as evenly as possible, three of each of the 100

per cent total, there remains at least 10 per cent with all four disabilities.
 [Cuthwellis, 1978, p. 9]

244. 6¼ minutes.
 [Cuthwellis, 1978, p. 9]

245. When the traveller crosses the International Date Line, which was internationally agreed with just such a purpose in mind, but only in 1884, long after the question had first troubled Carroll.
 [Carroll, 1961, p. 4]

246. Provided friction is neglected, the weight at the other end of the rope rises also, to match the monkey. Given friction in the pulley wheel, the weight will move up less than the monkey, or indeed not at all, if the pulley is sufficiently stiff.
 [Carroll, 1961, p. 268]

247. The answer is *always* £12 18s 11d, whatever the initial sum chosen.
 [Carroll, 1961, p. 269]

248. Assume, as is necessary but also implied, that when one basket is within reach of the window, the other is at ground level. Raise one basket, place the weight in it and lower the weight, raising the second basket, into which the son steps and descends to the ground.

 Lower the top basket, containing the weight. The son steps into this basket at ground level and the daughter descends in the other basket, raising the son and the weight.

 Lower the weight again; the son descends against it, and the daughter gets into the same basket as the son. The queen gets into the basket containing the weight and descends against the weight of son and daughter.

 The daughter steps out at the top, before the queen steps out of the bottom basket; the son remains and descends against the weight. The son steps out and the weight descends to the ground. The son gets in, and the daughter descends against son and weight.

 Finally, the son gets out, the weight descends, and the son goes down in the other basket, against the weight.
 [Carroll, 1961, p. 318]

249. The amounts are equal. Moreover, it makes not the slightest difference whether the water and brandy were stirred a little or a lot, or whether the glasses contained equal quantities of liquid initially. The quantity of liquid in each glass at the end is the same as at the start, and therefore what one has lost, the other has gained.

[Hudson, 1954, Appendix A]

250. 'A level mile takes $\frac{1}{4}$ hour, up hill $\frac{1}{3}$, down hill $\frac{1}{6}$. Hence to go and return over the same mile, whether on the level or on the hill-side, takes $\frac{1}{2}$ an hour. Hence in 6 hours they went 12 miles out and 12 back. If the 12 miles out had been nearly all level, they would have taken a little over 3 hours; if nearly all up hill, a little under 4. Hence $3\frac{1}{2}$ hours must be within $\frac{1}{2}$ an hour of the time taken in reaching the peak; thus, as they started at 3, they got there within $\frac{1}{2}$ an hour of $\frac{1}{2}$ past 6.'

[Carroll, 1958, p. 77]

251. For simplicity of calculation, turn the amounts given into modern pence, so that the customer has 120, 24 and 6; the shopkeeper has 60, 12 and 1; and the friend has 48, 30, 4 and 3.

Then the customer gives 120 + 6 to the shopkeeper and 24 to the friend; the shopkeeper gives 60 to the customer and 12 + 1 to his friend; and the friend gives 30 + 4 to the shopkeeper, and 3 to the customer. On balance the friend has gained or lost nothing, and the shopkeeper is 87d better off, so the customer has given the shopkeeper 7s 3d.

[Carroll, 1961, p. 317]

252.

[Cuthwellis, 1978, p. 14]

253. 'The cat wins, of course. It has to make precisely 100 leaps to complete the distance and return. The dog, on the contrary, is compelled to go 102 feet and back. Its thirty-third leap takes it to the 99-foot mark and so another leap, carrying it two feet beyond the mark, becomes necessary. In all, the dog must make 68 leaps to go the distance. But it jumps only two-thirds as quickly as the cat, so that while the cat is making 100 leaps the dog cannot quite make 67.

'But Barnum had an April Fool possibility up his sleeve. Suppose that the cat is named Sir Thomas, and the dog is female! The phrase "she makes three leaps to his two" would then mean that the dog would leap 9 feet while the cat went 4. Thus when the dog finishes the race in 68 leaps, the cat will have travelled only 90 feet and 8 inches.'

[Loyd, 1959, Book 1, No. 14]

254.

$$\begin{array}{r} 853 \\ 749\overline{)638897} \\ 5992 \\ \hline 3969 \\ 3745 \\ \hline 2247 \\ 2247 \end{array}$$

[Loyd, 1959, Book 1, No. 41]

255. 'In this remarkable problem we find that the lake contained exactly 11 acres, therefore the approximate answer of "nearly 11 acres" is not sufficiently correct. This definite answer is worked out by the Pythagorean law, which proves that on any right-angle triangle the square of the longest side is equal to the sum of the squares of the other two sides.

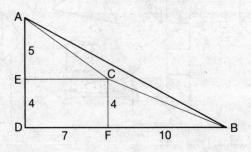

'In the illustration ABD represents our triangle, AD being 9 units long and BD 17, because 9 × 9 equals 81, which added to 17 × 17 (289) equals the 370 acres of the largest field. AEC is a right-angle triangle, and the square of 5 (25) added to the square of 7 (49) shows that the square on AC equals 74. CBF also is a right-angle triangle. The square of its sides, 4 and 10, prove the square estate on BC to equal 116 acres. The area of our triangle ADB is clearly half of 9 × 17, which equals 76.5 acres. Since the areas of the oblong and two triangles can plainly be seen to be 65.5, we deduct the same from 76.5 to prove that the lake contains exactly 11 acres.'

[Loyd, 1959, Book 1, No. 36]

256. 'The black pieces of paper are nothing but a delusion and a snare. The pieces are placed to make a little white horse in the centre as shown.

'It was this trick of the White Horse of Uffington which popularized the slang expression, "Oh, but that is a horse of another colour!"'

[Loyd, 1959, Book 1, No. 45]

257.

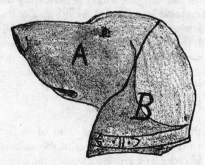

[Loyd, 1954, Book 1, No. 102]

258. Move the pieces in this sequence: 14 11 12 8 7 6 10 12 8 7 4 3 6 4 7 14 11 15 13 9 12 8 4 10 8 4 14 11 15 13 9 12 4 8 5 4 8 9 13 14 10 6 2 1. This is forty-four moves, the minimum possible.

The original puzzle is impossible to solve. All the possible positions in which the fifteen titles and the single space might be arranged, can be divided into two equal classes, call them the 'odd' and the 'even' positions. From an 'odd' it is possible to reach any other 'odd' position, but quite impossible to reach any 'even' position.

In particular, if in any particular position one pair of tiles is swopped, the new position can never be obtained from the original position. If, however, two such swops are made, the position is obtainable.

[Loyd, 1959, Book 1, No. 21]

259. The boy is five years old.
[Loyd, 1959, Book 2, No. 86]

260. Five odd 'figures' will add up to 14 as follows:

$$11$$
$$1$$
$$1$$
$$1$$
$$\overline{}$$
$$14$$

[Loyd, 1959, Book 2, No. 69]

261. If we let x be the bridge's length in feet, then the cow stands $\frac{1}{2}x - 5$ from one end and $\frac{1}{2}x + 5$ from the other. The train is $2x$ from the nearest end. The cow can travel $(\frac{1}{2}x - 5) + (\frac{1}{2}x + 4\frac{3}{4})$ in the same time that the train travels $(2x - 1) + (3x - \frac{1}{4})$. These two periods of time reduce to $(x - \frac{1}{4})$ and $5(x - \frac{1}{4})$, so we see that the train is five times faster than the cow. With this information we write the equation:

$$2x - 1 = 5(\tfrac{1}{2}x - 5)$$

This gives x, the length of the bridge, a value of 48 feet. The actual speed of the train plays no part whatever in this calculation, but we need to know it in order to learn the speed of the cow. Since we are told that the train travelled at 90 miles per hour, we know the cow's gait to be 18 miles per hour.

[Loyd, 1959, Book 2, No. 166]

262. 'The cheapest way to make an endless chain out of the six five-link pieces is to open up all five links of one piece, then use them for joining the remaining five pieces into an endless chain. The cost of this would be $1.30, which is 20 cents cheaper than the cost of a new endless chain.'

[Loyd, 1959, Book 2, No. 25]

263.

[Loyd, 1959, Book 1, No. 51]

264. 'Let 1 be the length of the army and the time it takes the army to march its length. The army's speed will also be 1. Let x be the total distance travelled by the courier and also his speed. On the courier's forward trip, his speed relative to the moving army will be $x - 1$. On the return trip his speed relative to the army will be $x + 1$. Each trip is a distance of 1 (relative to the army), and the two trips are completed in unit time, so we can write the following equation

$$\frac{1}{x - 1} + \frac{1}{x + 1} = 1$$

This can be expressed as the quadratic: $x^2 - 2x - 1 = 0$, for which x has the positive value of $1 + \sqrt{2}$. We multiply this by 50 to get the final answer of 120.7 miles.

'[In Part II] the courier's speed relative to the moving army is $x - 1$ on his forward trip, $x + 1$ on his backward trip, and $\sqrt{x^2 - 1}$ on his two diagonal trips. (It does not matter where he starts his round trip, so to simplify the problem we think of him as starting at a rear corner of the square instead of at the centre of the rear.) As before, each trip is a distance of 1 relative to the army, and since he completes the four trips in unit time we can write:

$$\frac{1}{x-1} + \frac{1}{x+1} + \frac{2}{\sqrt{x^2-1}} = 1$$

This can be expressed as the fourth degree equation: $x^4 - 4x^3 - 2x^2 + 4x + 5 = 0$, which has only one root that fits the problem's conditions: 4.18112. This is multiplied by 50 to get the final answer of 209.056 miles.'

[Loyd, 1959, Book 2, No. 146]

265. Martin Gardner comments, 'Loyd's *Cyclopaedia* does not answer this difficult problem ... The best procedure, supported by the answers to similar problems in Dudeney's puzzle books, seems to be the following:

'C, the slowest walker, always rides the tandem. He and A, the fastest walker, ride the tandem for 31.04 miles while B is walking. A dismounts, and C turns around and rides back to pick up B at a spot 5.63 miles from the start. B and C remain on the bicycle for the remainder of the journey, arriving at the same time that A arrives on foot. The total time is a little less than 2.3 hours.'

[Loyd, 1959, Book 2, No. 123]

266.

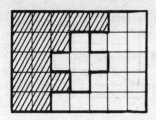
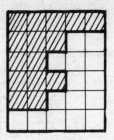

[Loyd, 1960, problem 144]

267.

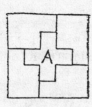

[Loyd (Jnr), 1928, p. 33]

268.

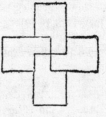

[Loyd (Jnr), 1928, p. 17]

269.

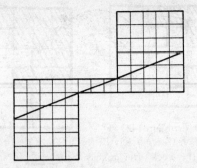

[Loyd (Jnr), 1928, p. 19]

270. Call the two 10-gallon cans A and B. This is Loyd's solution, in tabular form:

A	B	5-qt	4-qt
10	10	–	–
5	10	5	–
5	10	1	4
9	10	1	–
9	10	–	1
4	10	5	1
4	10	2	4
8	10	2	–
8	6	2	4
10	6	2	2

[Loyd (Jnr), 1928, p. 21]

271. The landlady placed 20 cents on the counter to pay for $1\frac{1}{2}$ pounds of bologna. Louis cut $2\frac{1}{4}$ pounds. She took $1\frac{1}{8}$ pounds for 15 cents and invested the remaining 5 cents in pickles.

[Loyd (Jnr), 1928, p. 25]

272. Seven pieces can be cut, though one of them is minute:

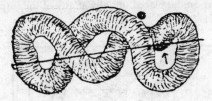

[Loyd (Jnr), 1928, p. 26]

273. 'On Sunday, the first day of the week, Kate promised to marry Danny "when the week after next is the week before last". Therefore she will marry Danny in 28 days after her promise. Had she promised a day earlier, then on Sunday, 22 days later, her promise would have fallen due.'

[Loyd (Jnr), 1928, p. 27]

274.

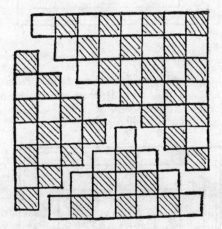

[Loyd (Jnr), 1928, p. 29]

275. 'The girl weighed $111\frac{1}{9}$ pounds when she arrived. She ate $1\frac{1}{9}$ pounds of breakfast food and gathered 10 pounds of samples, which increased her weight by 10 per cent.'

[Loyd (Jnr), 1928, p. 33]

276. The historical incident was 'the dropping of the tea into the sea', otherwise known as the Boston Tea Party.

[Loyd (Jnr), 1928, p. 47]

277. The Queen can be placed initially at any of the turning points in the path.

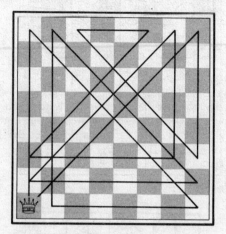

[White, 1913, p. 42]

278. This is Loyd's solution, from *Sam Loyd's Puzzle Magazine*, April 1908. A Queen's Tour from d1 (the white queen's initial square in a game of chess) is impossible without resorting to non-chess moves.

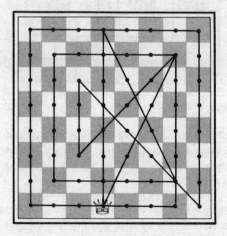

279.

[White, 1913, p. 52]

280.

[White, 1913, p. 52]

281. (a) 1 c4 c5; 2 Qa4 Qa5; 3 Qc6 Qc3; 4 Q × B mate. *Or* 1 d4 d5;
2 Qd3 Qd6; 3 Qh3 Qh6; 4 Q × B mate.

(b) 1 e4; 2 Ke2; 3 Ke3; 4 Qf3; 5 Ne2; 6 b3; 7 Ba3; 8 Nd4+,
e × d mate.

(c) 1 f3 e5; 2 Kf2 h5; 3 Kg3 h4+; 4 Kg4 d5 mate.

(d) 1 e3 a5; 2 Qh5 Ra6; 3 Q × a5 h5; 4 Q × c7 Rah6; 5 h4 f6;
6 Q × d7+ Kf7; 7 Q × b7 Qd6; 8 Q × b8 Qh2; 9 Q × c8
Kg6; 10 Qe6 stalemate.

(e) 1 f4 e5; 2 Kf2 Qf6; 3 Kg3 and Black can force perpetual check.

[White, 1913, pp. 58–9]

282. 'The illustration will show that the triangular piece of cloth may be cut into four pieces that will fit together and form a perfect square. Bisect AB in D and BC in E; produce the line AE to F making EF equal to EB; bisect AF in G and describe the arc AHF; produce EB to H, and EH is the length of the side of the required square; from E with distance EH, describe the arc HJ, and make JK equal to BE; now, from the points D and K drop perpendiculars on EJ to L and M . . .

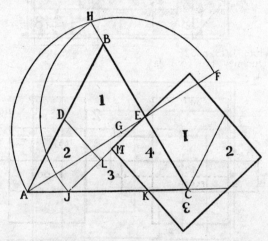

'I exhibited this problem before the Royal Society, at Burlington House, on 17th March 1905, and also at the Royal Institution in the following month, in the more general form: "A New Problem on Superposition: a demonstration that an equilateral triangle can be cut into four pieces that may be reassembled to form a square, with some examples of a general method for transforming all rectilinear figures into squares by dissection."

'I add an illustration showing the puzzle in a rather curious practical form, as it was made in polished mahogany with brass hinges for use by certain audiences. It will be seen that the four pieces form a sort of chain, and then when they are closed up in one direction they form the triangle, and when closed in the other direction they form the square.'

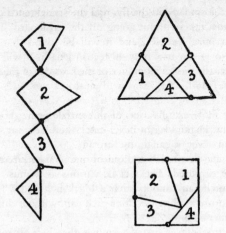

[Dudeney, 1907, No. 26]

283.

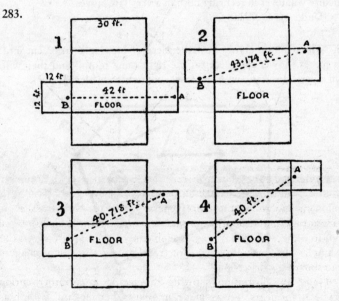

'Imagine the room to be a cardboard box. Then the box may be cut in various ways, so that the cardboard may be laid on the table. I show four of these ways, and indicate in every case the relative positions of the spider and the fly, and the straightened course which the spider must take without going off the cardboard. These are the four most favourable cases, and it will be found that the shortest route is in No. 4, for it is only 40 feet in length. It will be seen that the spider actually passes along five of the six sides of the room!'

[Dudeney, 1907, No. 75]

284. 'A very short examination of this puzzle game should convince the reader that Hendrick can never catch the black hog, and that the white hog can never be caught by Katrun.

'Each hog merely runs in and out of one of the nearest corners and can never be captured. The fact is, curious as it must at first sight appear, a Dutchman cannot catch a black hog and a Dutchwoman can never capture a white one! But each can, without difficulty, catch one of the other colour.

'So if the first player just determines that he will send Hendrick after the white porker and Katrun after the black one, he will have no difficulty whatever in securing both in a very few moves.'

[Dudeney, 1907, No. 78]

285. 'The diagram shows how the piece of bunting is to be cut into two pieces. Lower the piece on the right one "tooth" and they will form a perfect square, with the roses symmetrically placed.'

[Dudeney, 1907, No. 77]

286. The diagram shows how seven of the planks are used to 'round off' the corner, so that the eighth plank can be used as a bridge to the other side.

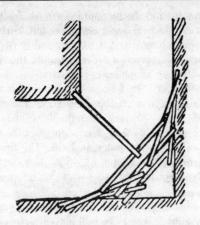

[Dudeney, 1907, No. 54]

287. This is the only possible arrangement. The casket is 20 inches square.

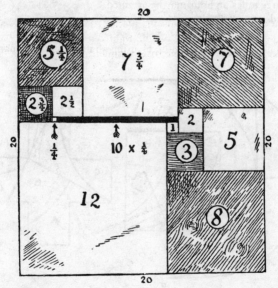

[Dudeney, 1907, No. 40]

288. This appears in *536 Puzzles and Curious Problems*, though another version, in which a dog runs back and forth between his master and an approaching friend, had appeared in *Modern Puzzles*. Question (1) is inserted merely in order to induce the reader to solve question (2) by adding up a long series of fractions.

The fly first meets car B in 1 hour 48 minutes. Since the cars are approaching each other at a combined speed of 150 miles an hour, they meet after $300/150 = 2$ hours, when the fly will be crushed.

(During that time the fly has been continuously flying in one direction or the other, at 150 miles an hour. The fly has therefore flown a total distance of 300 miles.)

[Dudeney, 1967, p. 26]

289. 'No doubt some of my readers will smile at the statement that a man in a boat on smooth water can pull himself across with the tiller rope! But it is a fact. If the jester had fastened the end of his rope to the stern of the boat and then, while standing in the bows, had given a series of violent jerks, the boat would have been propelled forward. This has often been put to practical test, and it is said that a speed of two or three miles an hour may be attained.'

[Dudeney, 1907, No. 52]

290. This is the simplest solution, by using an intermediate square.

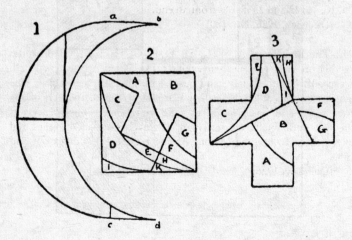

The crescent is in the form of two equal straight lines, *a* and *b*, joined by two identical circular arcs. Making the cuts in the first figure, the four pieces will form the square in the second, which is then dissected by the dotted lines into the Greek Cross, with a total of ten pieces.

[Dudeney, 1907, No. 37]

291. This puzzle is typical of Dudeney's interest in properties of numbers. 1,111,111 has only two factors, apart from itself and unity:

$$1,111,111 = 239 \times 4649, \text{ both factors being prime.}$$

So, either 239 cats caught 4649 mice each, or 4649 cats caught 239 mice each.

[Dudeney, 1907, No. 47]

292. 'The first player must place his first cigar *on end* in the exact centre of the table. Now, whatever the second player may do throughout, the first player must always repeat it in an exactly diametrically opposite position.' In this way the first player can be certain of always placing the last cigar.

[Dudeney, 1917, No. 398]

293. Dudeney placed the marks at 1, 4, 5, 14, 16, 23, 25 and 31 inches from one end. He also gave another solution, with the marks at 1, 2, 3, 4, 10, 16, 22 and 28 inches from an end.

[Dudeney, 1926, No. 180]

294. 'The nine men, A, B, C, D, E, F, G, H, J, all go 40 miles together on the 1 gall. in their engine tanks, when A transfers 1 gall. to each of the other eight, and has 1 gall. left to return home. The eight go another 40 miles, when B transfers 1 gall. to each of the other seven and has 2 galls. to take him home.' This process is repeated, until finally the last man, J, travels another 40 miles and has 9 gallons to take him home, having travelled 360 miles out and home.

[Dudeney, 1926, No. 49]

295.

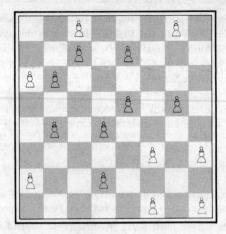

[Dudeney, 1917, No. 317]

296. With a choice of 125 or 100 yards, it is possible to go round in just twenty-six strokes: the strokes required for all the holes are evident, with the exception of hole 5, which is reached in three drives of 125 yards and one stroke backwards of 100 yards.

[Dudeney, 1907, No. 32]

297. This is a variant of the Bridges of Königsberg. There are just two cells with an odd number of doors, and, in order to pass through every door just once, it is necessary to start at one of these cells and end at the other. Therefore, the route must start at the starred 'odd' cell on the outside, and the fair demoiselle is at the other starred cell.

[Dudeney, 1907, No. 34]

298. 'Add together the ten weights and divide by 4 and we get 289 lbs, as the weight of the five trusses together. If we call the five trusses in the order of weight, A, B, C, D and E, the lightest being A and the heaviest E, then the lightest 110 lbs must be the weight of A and B, and the next lightest, 112 lbs, the weight of A and C. Then the two heaviest, D and E, must weigh 121 lbs, and C and E must weigh 120 lbs. We thus know that A, B, D and E weigh together 231 lbs, which gives us the weight of C as 58 lbs. Now, by mere subtraction, we find the weights of the other five trusses, 54 lbs, 56 lbs, 59 lbs and 62 lbs, respectively.'

[Dudeney, 1917, No. 101]

299. This beautiful solution requires just five pieces and only two cuts. The distance AB is equal to one half of the hypotenuse of the triangle. The triangle should not be too large – Dudeney notes that if it is larger than the square in area, then a dissection requires six pieces.

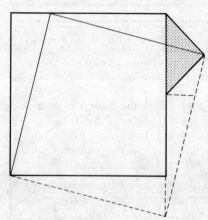

[Dudeney, 1917, No. 152]

300. Only sixteen moves are required, and these are the only two possible minimum solutions.

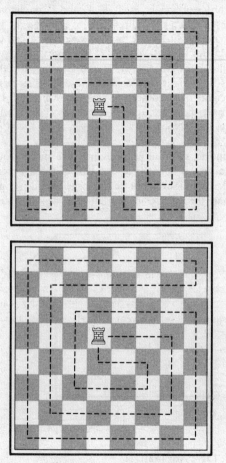

[Dudeney, 1917, No. 320]

301. Three coins are placed as on the left, each touching the others, and then two coins are added as in the second figure.

[Dudeney, 1917, No. 419]

302. De Morgan was born in 1806 and was 43 in the year $43^2 = 1849$ (see problem 231).

Jenkins was born in 1860, was $5^2 + 6^2 = 61$ in the year $5^4 + 6^4 = 1921$, and $2 \times 31 = 62$ in the year $2 \times 31^2 = 1922$, and $3 \times 5 = 15$ in the year $3 \times 5^4 = 1875$.

[Dudeney, 1926, No. 23]

303. N and $N/(N - 1)$ have this property, whatever the integral value of N, other than unity.

$$N + \frac{N}{N - 1} = N \times \left(1 + \frac{1}{N - 1}\right) = \frac{N^2}{N - 1}$$

[Dudeney, 1926, No. 93]

304. 'According to the conditions, in the strict sense in which one at first understands them, there is no possible solution to this puzzle. In such a dilemma one always has to look for some verbal quibble or trick. If the owner of house A will allow the water company to run their pipes for houses B and C through his property (and we are not bound to assume that he would object), then the difficulty is got over, as shown in our illustration.'

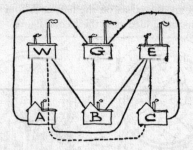

[Dudeney, 1917, No. 251]

305. Arrange the six pennies as in the first figure. This can be done exactly. Next move coin 6 as in the next figure. This is also exact. Finally slide out 5 and place it against 2 and 3, and move 3 to just touch 6 and 5.

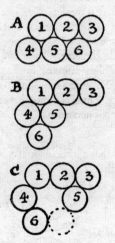

This puzzle is usually presented as the problem of merely transforming the first arrangement into the final arrangement, or sometimes of transforming a triangle of six pennies (move 4 in the first figure to touch 5 and 6) into the circle. Dudeney's presentation seems to me much superior.

[Dudeney, 1926, No. 213]

306. Thirteen coins can be placed as shown in this figure:

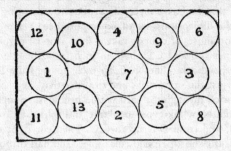

[Dudeney, 1917, No. 429]

307. 'The secret of the bun puzzle lies in the fact that, with the relative dimensions of the circles as given, the three diameters will form a right-angled triangle, as shown by A, B, C. It follows that the two smaller buns are exactly equal to the large bun. Therefore, if we give David and Edgar the two halves marked D and E, they will have their fair shares – one quarter of the confectionery each. Then if we place the small bun, H, on the top of the remaining one and trace its circumference in the manner shown, Fred's piece, F, will exactly equal Harry's small bun, H, with the addition of the piece marked G – half the rim of the other. Thus each boy gets an exactly equal share, and there are only five pieces necessary.'

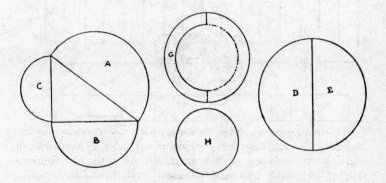

[Dudeney, 1917, No. 148]

308. 'The reader will probably feel rewarded for any care and patience that he may bestow on cutting out the cardboard chain. We will suppose that he has a piece of cardboard measuring 8 in. by $2\frac{1}{2}$ in., though the dimensions are of no importance. Yet if you want a long chain you must, of course, take a long strip of cardboard. First rule pencil lines B B and C C, half an inch from the edges, and also the short perpendicular lines half an inch apart. Rule lines on the other side in just the same way, and in order that they shall coincide it is well to prick through the card with a needle the points where the short lines end. Now take your penknife and split the card from A A down to B B, and from D D up to C C. Then cut right through the card along all the short perpendicular lines, and half through the card along the short portions of B B and C C that are not dotted. Next turn the card over and cut half through along the short lines on B B and C C at the places that are immediately beneath the dotted lines on the upper side. With a little careful separation of the parts with the penknife, the cardboard may now be divided into two interlacing ladder-like portions; and if you cut away all the shaded parts you will get the chain, cut solidly out of the cardboard, without any join.

'It is an interesting variant of the puzzle to cut out two keys on a ring – without join.'

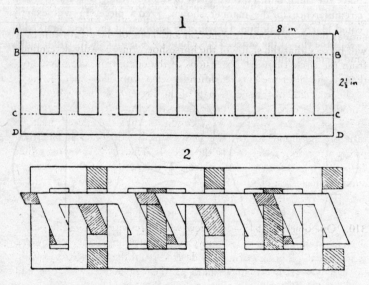

[Dudeney, 1917, No. 177]

309. 'The puzzle was to cut the two shoes (including the hoof contained within the outlines) into four pieces, two pieces each, that would fit together and form a perfect circle. It was also stipulated that all four pieces should be different in shape. As a matter of fact, it is a puzzle based on the principle contained in that curious Chinese symbol the Monad.

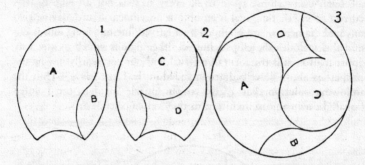

'The above diagrams give the correct solution to the problem. It will be noticed that 1 and 2 are cut into the required four pieces, all different in shape, that fit together and form the perfect circle shown in diagram 3. It will further be observed that the two pieces A and B of one shoe and the two pieces C and D of the other form two exactly similar halves of the circle – the Yin and Yang of the great Monad. It will be seen that the shape of the horseshoe is more easily determined from the circle than the dimensions of the circle from the horseshoe, though the latter presents no difficulty when you know that the curve of the long side of the shoe is part of the circumference of your circle. The difference between B and D is instructive, and the idea is useful in all such cases where it is a condition that the pieces must be different in shape. In forming D we simply add on a symmetrical piece, a curvilinear square, to the piece B. Therefore, in giving either B or D a quarter turn before placing in the new position, a precisely similar effect must be produced.'

[Dudeney, 1917, No. 160]

310. 'One object that I had in view when presenting this little puzzle was to point out the uncertainty of the meaning conveyed by the word "oval". Though originally derived from the Latin word *ovum*, an egg, yet what we understand as the egg-shape (with one end smaller than the other) is only one of many forms of the oval; while some eggs are spherical in shape, and a sphere or circle is most certainly not an oval. If we speak of an ellipse – a conical ellipse – we are on safer ground, but here we must be careful of error. I recollect a Liverpool town councillor, many years ago, whose ignorance of the poultry-yard led him to substitute the word "hen" for "fowl", remarking, "We must remember, gentlemen, that although every cock is a hen, every hen is not a cock!" Similarly, we must always note that although every ellipse is an oval, every oval is not an ellipse. It is correct to say that an oval is an oblong curvilinear figure, having two unequal diameters, and bounded by a curved line returning into itself; and this includes the ellipse, but all other figures which in any way approach towards the form of an oval without necessarily having the properties above described are included in the term "oval". Thus the following solution that I give to our puzzle involves the pointed "oval" known among architects as the "vesica piscis".

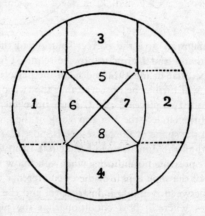

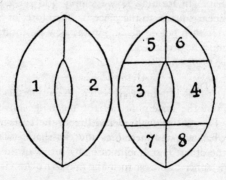

'The dotted lines in the table are given for greater clearness, the cuts being made along the other lines. It will be seen that the eight pieces form two stools of exactly the same size and shape with similar hand-holes. These holes are a trifle longer than those in the schoolmaster's stools, but they are much narrower and of considerably smaller area. Of course 5 and 6 can be cut out in one piece – also 7 and 8 – making only *six pieces* in all. But I wished to keep the same number as in the original story.

'When I first gave the above puzzle in a London newspaper, in competition, no correct solution was received, but an ingenious and neatly executed attempt by a man lying in a London infirmary was accompanied by the following note: "Having no compasses here, I was compelled to improvise with the aid of a small penknife, a bit of firewood from a bundle, a piece of tin from a toy engine, a tin tack, and two portions of a hairpin, for points. They are a fairly serviceable pair of compasses, and I shall keep them as a memento of your puzzle "'

[Dudeney, 1917, No. 157]

311. The sum is $9567 + 1085 = 10652$.
[*Strand Magazine*, July 1924]

312. 'The first step is to find the distance travelled by the spiders. We use the formula

$$Distance = Velocity \times Time.$$

According to our stated conditions, the velocity is 0.65 miles per hour; but we want to get this in inches per second, since the time in our problem is given in seconds and part of the distance in inches. Calculating:

$$\frac{0.65 \text{ (mile)}}{1 \text{ (hour)}} = \frac{0.65 \times 5280 \times 12}{60 \times 60} = \frac{41,184 \text{ (inches)}}{3600 \text{ (seconds)}}$$

gives us our velocity. Then

$$Distance = \frac{41,184}{3600} \times \frac{625}{11} = 650 \text{ inches}$$

That is how far each spider travelled. But we were asked to find the dimensions of the room. Therefore, now, let us spread out its walls, ceiling, and floor onto a plane, much as if we were to open out the six faces of a cardboard box to make one flat piece. Since there are

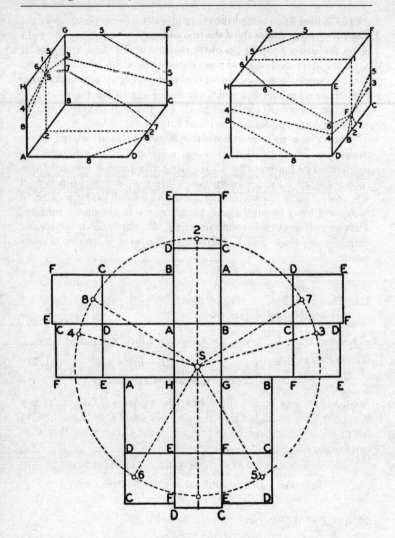

several possible paths, this must be done in all possible ways, keeping the wall from which the spiders start fixed, and laying down the others in such a way as to keep every face attached to another along a common edge (see diagram). We can then see that the following eight paths are possible:

1. Over the floor and parts of the two end walls.
2. Over the ceiling and parts of the two end walls.
3. Over one side wall and parts of the two end walls.
4. Over the other side wall and parts of the two end walls.
5. Over the ceiling and one side wall and parts of the two end walls.
6. Over the ceiling and the other side wall and parts of the two end walls.
7. Over the floor and one side wall and parts of the two end walls.
8. Over the floor and the other side wall and parts of the two end walls.

Denote the three dimensions of the room by l = length, w = width, and h = height. The lengths of the various paths may then be expressed:

$$Distance = l + h = 650 \text{ inches.} \qquad (1, 2)$$

$$Distance = \sqrt{160^2 + (l + w)^2} = 650 \text{ inches} \qquad (3, 4)$$

$$Distance = \sqrt{\left(\frac{h + w}{2} + 80\right)^2 + \left(l + \frac{h + w}{2} - 80\right)^2}$$

$$= 650 \text{ inches} \qquad (5, 6, 7, 8)$$

To eliminate the radical signs, we may rewrite these equations:

$$(l + h)^2 = 650^2 \qquad (1, 2)$$

$$160^2 + (l + w)^2 = 650^2 \qquad (3, 4)$$

$$\left(\frac{h + w}{2} + 80\right)^2 + \left(l + \frac{h + w}{2} - 80\right)^2 = 650^2 \qquad (5, 6, 7, 8)$$

From these we find:

$$l + h = 650 \qquad (I)$$

$$(l + w)^2 = 650^2 - 160^2 = 396,900 = 630^2$$

$$l + w = 630 \qquad (II)$$

Adding (I) and (II), we have

$$2l + h + w = 1280$$

whence

$$l + \frac{h + w}{2} = 640.$$

The third equation now yields

$$\left(\frac{h + w}{2} + 80\right)^2 = 650^2 - (640 - 80)^2 = 650^2 - 560^2$$

$$= 108,900 = 330^2$$

whence $$h + w = 500 \qquad \text{(III)}$$

From half the sum of equations (I), (II) and (III), we get

$$l + h + w = 890$$

from which we obtain, by subtracting (III), (I) and (II) in turn,

$$l = 390, w = 240, h = 260'$$

[Kraitchik, 1955, pp. 18–19]

313. This is essentially the only solution. Note the rough symmetry about the central column when the hexagon is in this orientation.

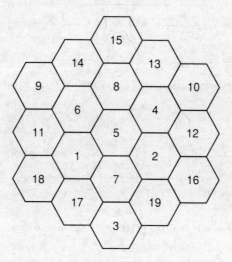

[Gardner, 1971, pp. 22–4]

314. The level fell. As long as it was in the boat, the cannon displaced its own *weight* of water. After it sank, it displaced only its own *volume*, which is a smaller quantity of water.

[After Williams and Savage, 1946]

315. The sum can change because the three numbers at the vertices, and no others, are counted twice.

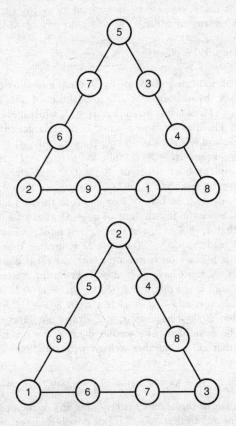

[Perelman, 1979, No. 49]

316. The front wheels are usually smaller in diameter than the back, and therefore turn a larger number of times, and wear and tear on the front axle is greater.

[Perelman, 1979, No. 65]

317. No it is not, because the single small cube at the centre of the original large cube has six faces, and a separate slice is needed to create each of those faces. Therefore six cuts is a minimum.

[Perelman, 1979, No. 122]

318. They had counted the same. Tom, as he walks to and fro, will meet some passers-by sooner, because he is approaching them, and others with a delay, because he is walking away from them, but by the time he returns to Mike, everyone who passed Mike will have passed Tom also.

[Perelman, 1979, No. 4]

319. 'We shall indicate the divisor by D, the quotient by Q, the digits of the divisor by $d_1, d_2, \ldots, d_7$, and those of the quotient by $q_1, q_2, \ldots, q_{10}$. Hence, it is given that $q_4 = 7$, while it is clear at once that $q_8 = 0$. The digits q_3 and q_7 must both be larger than q_4 (hence larger than 7) and less than q_2 and q_6; consequently, $q_3 = q_7 = 8$ and $q_2 = q_6 = 9$. From $q_7 = 8$ it follows that $8 \times D$ is less than 10,000,000 and at least equal to $10,000,000 - 97,999 = 9,902,001$; hence D must be less than 1,250,000 and greater than 1,237,750, so that $\mathbf{d_1 = 1}$, $\mathbf{d_2 = 2}$, and $d_3 = 3$ or 4. From this it follows further that $q_5 = 8$. Since the fourth digit of $q_5 \times D$, thus of $8 \times D$, is a 7, this shows that $d_4 = 4$ or 9, and that d_6 is at most 4. The assumption that $d_3 = 3$ leads to $d_4 = 9$ (because D is greater than 1,237,750), from which it follows (in connection with the third digit of $q_9 \times D$ being 7) that we must have $q_9 = 2$ or 7. From the thirteenth row of the division sum it is evident that $(800 + q_9 + 1) \times D$ is a ten-digit number. However, $803 \times d_6 d_7$ is less than $803 \times 1,240,000$, hence less than 995,720,000, so that $q_9 = 2$ drops out, and only $q_9 = 7$ remains to be examined. The second digit of the product obtained when $7q_{10}$ (that is, the number written with the digits 7 and q_{10}) is multiplied by D is a 7; but we have

$$7q_{10} \times 1,239,7d_6 d_7 = 86,779,000 + (q_{10} \times 1,239,7d_6 d_7) + (70 \times d_6 d_7)$$

the second digit of this number is not 7 for any of the possible values $1, 2, \ldots, 8$ of q_{10}, so that $q_9 = 7$ is not possible, either, which makes $d_3 = 3$ drop out. So we must have $\mathbf{d_3 = 4}$, and $D = 1,24d_4,7d_6 d_7$ (where $d_4 = 4$ or 9). The third digit of $q_9 \times 1,249,7d_6 d_7$ is not a 7 for any of the possible values of q_9, so we must have $\mathbf{d_4 = 4}$. From the fact that the third digit of $q_9 \times 1,244,7d_6 d_7$ is a 7 it follows that $q_9 = 4$. The second digit of

$$4q_{10} \times 1,244,7d_6 d_7 = 49,788,000 + (q_{10} \times 1,244,7d_6 d_7) + (40 \times d_6 d_7)$$

is a 7, from which it follows that $q_{10} = 6$. The seventh digit of $898,046 \times 1,244,7d_6 d_7 = 1,117,797,856,200 + 898,046 \times d_6 d_7$ is a 7, from which we can deduce (since $d_6 d_7$ is less than 50) that $d_6 d_7 =$

$k \times 11$, where $k = 0, 1, 2, 3,$ or 4. Hence, for the dividend $Q \times D$ we find:

$$q_1, 987,898,046 \times 1,244,7d_6d_7 = (q_1 \times 1,244,700,000,000,000)$$
$$+ 1,229,636,697,856,200 + (k \times 10,866,878,605)$$
$$+ (q_1 \times k \times 11,000,000,000)$$

The seventh digit of this number is the last digit of $6 + k \times (q_1 + 1)$. This has to be a 7, by the terms of the problem, so that $k \times (q_1 + 1)$ must end in 1. In connection with $k = 0, 1, 2, 3,$ or 4 (keeping in mind that q_1 is different from 0), it follows that $k = 3$, hence $d_6 = d_7 = 3$, and $q_1 = 6$. Consequently

$$D = 1,244,733 \qquad Q = 6,987,898,046$$

The dividend is then found as the product

$$Q \times D = 8,698,067,298,491,718'$$

[Schuh, 1968, pp. 319–20]

320. Arranged as in the diagram, the top brick overhangs the second brick by one half of a brick length. The centre of gravity of the two

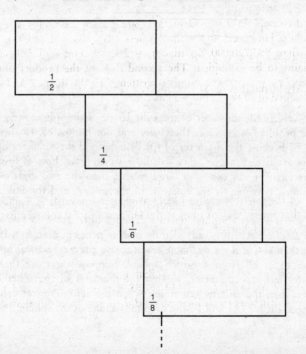

bricks together is at G, so that they can together overhang the third brick by $\frac{1}{4}$ brick length. The first three can overhang the fourth by $\frac{1}{6}$ brick length and so on.

The total possible overhang is the sum of the series:

$$\tfrac{1}{2} + \tfrac{1}{4} + \tfrac{1}{6} + \tfrac{1}{8} + \tfrac{1}{10} + \ldots$$

which increases without limit, being just one half of the harmonic series $1 + \frac{1}{2} + \frac{1}{3} + \frac{1}{4} \ldots$

With just four bricks, the maximum overhang is $\frac{1}{2} + \frac{1}{4} + \frac{1}{6} + \frac{1}{8} = 1\frac{1}{24}$, so that the top brick completely overhangs the bottom brick.

If the bricks are replaced by a pack of fifty-two cards, the maximum overhang is a little more than $2\frac{1}{4}$ card lengths.

321. With the method of the last puzzle, an overhang of at most $1\frac{1}{24}$ is possible with four bricks. In the following arrangement the overhang is $(15 - 4\sqrt{2})/8$, or a little over 7/6, and thus a little more than in the previous puzzle.

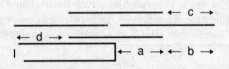

[Ainley, 1977, p. 8]

322. My birthday is on 31 December, and today is 1 January.

323. Neither the diameter of the cylinder or of the sphere are given. If the problem is genuine, therefore, and not impossible to solve, it must be because the diameter of the cylinder and the diameter of the sphere are irrelevant. In other words, if a cylindrical hole of length 6 inches is to be cut out of a large sphere, then the two ends of the hole must be close to an 'equator' of the sphere, and the hole very wide, the small amount of material left will be equal in volume to that left if a very thin and narrow cylinder of length 6 inches is cut from a sphere which is only slightly over 6 inches in diameter. If this is so, then you will get the same result if you cut a cylindrical hole of no width at all from a 6-inch sphere. Suppose therefore that the cylinder has zero diameter, so that the diameter of the sphere is 6 inches, then the volume remaining when the zero-diameter cylinder has been drilled out will be the original volume of the sphere, $\frac{4}{3}\pi 3^3 = 36\pi$.

This is in fact the correct answer, and is independent of the measurements that were omitted.

There is a matching 'surprise' in two dimensions. If the longest line segment that can be drawn in the space between two concentric circles, without crossing the inner circle, is 6 inches long, what is the area of the annulus? The area is $\pi \times 3^2 = 9\pi$, regardless of the actual dimensions of the circles.

324. The suggestion that the sums named ought to add up to £30 is nonsense. The diners have spent a total of £27, of which £25 was for the dinner, and £2 for the waiter.

325. On submersion, the lead and the iron weights will be supported by the water, according to the quantity of water they displace. Since iron weighs less than lead, the iron equal in weight to the lead will displace more water. The iron will therefore rise and the lead sink.

326. 'If the slice is a plane one, then since the top and bottom faces of the original cube are parallel, the two lines *ac* and *bd* will be parallel. However, to discover whether two lines are parallel in a perspective drawing is not trivial, so a little more cunning is required.

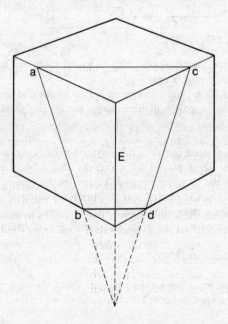

'Reconstruct the missing edge, *E*. Then *ab* will meet *E*, because both lines lie in one face of the cube. Similarly, *cd* meets *E*. IF *abcd* is a plane slice, then *ab* and *cd* will meet. In this case all three lines must meet in the same point, because all three lines do not lie in the same plane.

'Checking in the solution diagram, they do indeed *appear* to meet in the same point and so the slice *could* be a plane one. If it is not, it is only because its surface is not plane although its edges lie in one plane.'

[Wells, 1979, problem 60]

327. One or other, or both, of the governments of Monia and Moria, and therefore, indirectly, their tax-payers. Should you find yourself in the fortunate position of this young man, you would indeed find that 'money grew on trees' as long as the contradictory exchange rates continued.

[After Northrop, 1960, pp. 8–9]

328. Only twenty-three people need be in the room, a surprisingly small number. The probability that there will not be two matching birthdays is then, ignoring leap years,

$$\frac{365 \times 364 \times 363 \times \ldots \times 343}{365^{23}}$$

which is approximately 0.493. This is less than $\frac{1}{2}$, and therefore the probability that a pair occurs is greater than 50–50.

With as few as fourteen people in the room the chances are better than 50–50 that a pair will have birthdays on the same day or on consecutive days.

329. Mary suggested that she take white in one game and black in the other. She then played as white the moves that were played against her by the master taking white, and for her replies in the same game, she played the moves that the master playing black made to her. The masters thus played each other, in effect, and in between laughing managed to play a decisive game, giving Mary her 50 per cent score.

[After Kraitchik, 1955]

330. The man must have shot the bear from the North Pole. The bear was therefore a polar bear, and its colour was white.

Phillips's original problem was 'Polar conundrum: Starting from the North Pole, I walk 40 miles due south, and then 30 miles due west. How far am I now from the North Pole?' The deliberate choice of 30 and 40 miles suggests a 3–4–5 right-angled triangle and the mistaken answer 50 miles.

331. Mrs Agabegyun *might* live at the North Pole. However, it is also possible that she is living in Antarctica, rather near the South Pole. Walking 5 miles south, she is very close to the pole, so near that her 5 mile walk in an easterly direction takes her a whole number of times round the South Pole, bringing her back to the point she reached by walking 5 miles south: the final 5 miles north returns her to base.

In this case, there are an infinite number of solutions, because the 5 mile journey east may take her once round the South Pole, twice, three times . . . !

332. One red card and one blue one. It is not possible to say, however, whether the red card or the blue was shown first.
[After Phillips, 1937, problem 10]

333. The boy with the dirty face saw his companion with a clean face, and suspected nothing. His companion, seeing the other with a grimy face, assumed that his face was also dirty, and went to wash it.
[Phillips, 1932, 'Time Tests of Intelligence', problem 12]

334. He sees that the other Wise Men are both laughing. But if they could see that his face was clean, they would realize at once on looking at each other that they were victims themselves. Since they do not, he must suppose that each is laughing at him. Therefore his forehead is marked too.

335. Four rungs remain submerged, because the ladder rises with the ship, which rises with the tide.
[Phillips, 1936, problem 0.1]

336. The engine-driver is Brown. Mr R lives at Leeds and either Mr B or Mr J in London. One third of Mr J's income could not be the guard's. So the guard's neighbour must be Mr B. So Mr J lives in London and is therefore the guard's namesake, i.e., the guard is J. Now B is not the fireman and so must be the engine-driver.
['Puzzle Pages Editorial', *Games and Puzzles* magazine, Nos. 33–4, Feb–March 1975]

337. 'Stritebatt's average of 30 is obtained by dividing his total of runs scored by the number of times he was Out. Let him be Out m times, and let him be Not Out n times. Then $12n$ runs more would give him an average of 35, i.e. 5 more, and $12n = 5m$. It follows that the number of completed innings is 17 or some multiple thereof, i.e. 17, 34, 51, etc. But if Stritebatt has 34 innings, he scores on 32 occasions, and since his lowest score is 17 and no two scores are the same, his lowest possible scores are:

$$17 \quad 18 \quad 19 \quad 20 \ldots 48$$

and since he is Not Out 10 times, his average must be at least 43.3. If he has 51 innings, his average must be higher. Hence he has 17 innings only and his scores are: 0, 0, 17, 18, 19 ... 31. (For 360, the total number of runs scored, can only be made up in this way.) His best score, therefore, is 31 Not Out.'

[Phillips, 1934, problem 18]

338. '1. There are five lodgers; hence the number of rashers originally on the dish must be 5, or 10, or 15, etc.

2. The number is 5. For the greatest possible original number would be reached in the following way: Smith takes $\frac{1}{2}$ rasher; Jones takes 1, total $\frac{3}{2}$; Evans takes 3, total $\frac{9}{2}$; and then either:

(a) Brown takes $\frac{13}{6}$, total $\frac{20}{3}$; Robinson takes $\frac{5}{3}$, total $\frac{25}{3}$.

(b) Robinson takes $\frac{3}{2}$, total 6; Brown takes $\frac{17}{8}$, total $\frac{65}{8}$.

'Both these are less than 10; hence the original number of rashers is five.

3. Since Evans always leaves at least one rasher he cannot immediately precede Smith; so there are three possibilities:

(a) Jones precedes Smith and takes 1 rasher; total $\frac{3}{2}$.

(b) Brown precedes Smith and takes $\frac{3}{2}$ rashers; total 2.

'In both these Evans must take three rashers, leaving only $\frac{1}{2}$ or 0 for the other two; this is impossible.

(c) Robinson precedes Smith and takes $\frac{1}{2}$ rasher; total 1.

'Evans must precede Robinson; for if not he must take 3 rashers, Jones must take 1, and there will be none left for Brown. Let Evans take x rashers, then either:

(1) Jones takes 1 rasher, total $x + 2$; Brown takes $\frac{x}{4} + \frac{9}{8}$, total $\frac{5x}{4} + \frac{25}{8}$.

'For this to be 5 we must have $x = \frac{3}{2}$.

(2) Brown takes $\frac{x}{3} + 1$, total $\frac{4x}{3} + 2$; Jones takes 1, total $\frac{4x}{3} + 3$.

'For this to be 5 we must have $x = \frac{3}{2}$.

'Hence Evans takes one and half rashers of bacon.'

[Phillips, 1934, problem 20, contributed by J. W. Frame]

339. 'One of the four quoted statements is true.

'(1) Let the first statement be true. Then we have two hatters (Mr D and Mr B). So this hypothesis is "out".

(2) Let the second statement be true. Then Mr G is d; Mr B is h; Mr H is b. It follows that Mr D is g.

(3) Let the third statement be true. Then Mr H is b. So Mr D is neither b, h nor d, and, once again, must be g.

(4) Let the fourth statement be true. Now Mr B is h, Mr G is not h, d or g, and so is b. Whence, as before, Mr D is g.

'Hence while we cannot with certainty identify any one of the others, we know that Mr Draper is the grocer.'

[Phillips, 1950, 'Inference' problems, No. 15]

340. 19 represents 25. If the statement of the problem is simplified, the answer is almost obvious: it states that $ab \times ab = cab$.

[Phillips, 1960, 'One Hundred Elementary Problems', problem 53]

341. (1) The raven's owner's feathered namesake must be a light-coloured bird. Hence the raven is owned by one of the following: Mr Dove, Mr Canary, Mr Gull, Mr Parrot. The first two of these are bachelors and the raven is owned by Mr Gull's wife's sister's husband – i.e. *Mr Parrot owns the raven.*

(2) Mr Crow owns a light-coloured bird, but Mr Crow's bird's human namesake is married. Hence Mr Crow owns either the parrot or the gull. But Mr Crow cannot own the parrot, for the parrot's owner's feathered namesake is owned by the human namesake of Mr Crow's bird; and Mr Parrot, we know, owns the raven. Therefore *Mr Crow owns the gull.*

(3) Mr Raven must own the parrot, the gull or the dove. But Mr Crow owns the gull, and if Mr Raven owns the parrot two people would own the raven. Whence *Mr Raven owns the dove* and

(4) *Mr Dove owns the canary.*

(5) The crow's owner is unmarried; hence *Mr Canary owns the crow.* Whence:

(6) *Mr Starling owns the parrot* and

(7) *Mr Gull owns the starling.*

[Phillips, 1960, 'Inference' problems, p. 165]

342. A flat tax of 25s for each window, plus 1s 6d for every square foot of window.

[Hadfield, 1939, 'Puzzles and Problems by Caliban', problem 3]

343. This is Eddington's own solution, as printed in Phillips, 1960.

Denote the four male animals by A_1, A_2, A_3, A_4 and the females by $-A_1$, $-A_2$, $-A_3$, $-A_4$.

A child's mistake can be described as a transformation, e.g.:

$$A_1, A_2, A_3, A_4 \rightarrow -A_1, -A_2, -A_3, -A_4$$

The transformation is equivalent to multiplying by a matrix; thus

$$\begin{bmatrix} -A_2 \\ A_3 \\ A_1 \\ A_1 \end{bmatrix} = \begin{bmatrix} 0 & -1 & 0 & 0 \\ 0 & 0 & 1 & 0 \\ 1 & 0 & 0 & 0 \\ 0 & 0 & 0 & -1 \end{bmatrix} \times \begin{bmatrix} A_1 \\ A_1 \\ A_2 \\ A_4 \end{bmatrix}$$

We denote the matrices representing the misidentifications of the n children by E_m ($m = 1, 2 \ldots n$) so that the transformation is written $A \rightarrow E_m A$.

Owing to the condition in the second paragraph of the problem the above transformation applies to the female as well as the male animals.

The condition in the third paragraph can now be written

$$E_m(E_p A) = -E_p(E_m)A$$

[m being not equal to p]

so that

$$E_p E_m = -E_m E_p \tag{1}$$

The condition in the fourth paragraph is

$$E_m{}^2 = 1 \text{ for boys} \qquad E_m{}^2 = -1 \text{ for girls.} \tag{2}$$

It has been proved that no more than five fourfold matrices can satisfy (1) and that (when the elements are real) three of them have positive squares and two negative squares. More usually the theorem is stated in the form that, with fourfold matrices, there cannot be more than five mutually anti-commuting square roots of -1, and that three of them are imaginary and two real.

An actual set satisfying (1) and (2) is:

Boys

$$\begin{bmatrix} 0 & 1 & 0 & 0 \\ 1 & 0 & 0 & 0 \\ 0 & 0 & 0 & 1 \\ 0 & 0 & 1 & 0 \end{bmatrix} \begin{bmatrix} 1 & 0 & 0 & 0 \\ 0 & -1 & 0 & 0 \\ 0 & 0 & 1 & 0 \\ 0 & 0 & 0 & -1 \end{bmatrix} \begin{bmatrix} 0 & 0 & 0 & 1 \\ 0 & 0 & -1 & 0 \\ 0 & -1 & 0 & 0 \\ 1 & 0 & 0 & 0 \end{bmatrix}$$

Girls

$$\begin{bmatrix} 0 & 1 & 0 & 0 \\ -1 & 0 & 0 & 0 \\ 0 & 0 & 0 & -1 \\ 0 & 0 & 1 & 0 \end{bmatrix} \begin{bmatrix} 0 & 0 & 0 & 1 \\ 0 & 0 & -1 & 0 \\ 0 & 1 & 0 & 0 \\ -1 & 0 & 0 & 0 \end{bmatrix}$$

A non-diagonal element signifies a mistake of species. Thus the only list with any names right is the second – a boy. He got two pairs right, but interchanged the sexes of the other two pairs.

The answer is therefore:
(1) *three nephews, two nieces*
(2) *a boy*
(3) *4 completely right; 4 wrong sex.*

(There are five other possible pentads besides the one given; but they all have the same characteristic, that there is only one diagonal matrix and it has two elements +1 and two −1. The sign of any of the five matrices can be reversed. The answer is unaffected by these variations.)

Appended is a specimen set of five lists which fulfil the conditions.

(T = Mr Tove, *t* = Mrs Tove, and so on)

Correct Names	1st Boy	2nd Boy	3rd Boy	1st Girl	2nd Girl
T	B	T	J	B	J
B	T	*b*	*r*	*t*	*r*
R	J	R	*b*	*j*	B
J	R	*j*	T	R	*t*
t	*b*	*t*	*j*	*b*	*j*
b	*t*	B	R	T	R
r	*j*	*r*	B	J	*b*
j	*r*	J	*t*	*r*	T

344. It makes two complete rotations.

345. No. Suppose that the chessboard is coloured, as is usual, with a checkerboard pattern of alternating black and white squares. The two squares removed from the opposite ends of a long diagonal will be of the same colour. Suppose that they are both white. Then the remaining squares are thirty-two black and only thirty white.

But when a domino is placed to cover a pair of adjacent squares, it will inevitably cover one black square and one white square. Therefore thirty-one dominoes can only cover thirty-one squares of each colour.

The removal of any pair of squares of the same colour makes a covering impossible. However, if the squares removed are of different colours, then a covering is always possible, as Gomory proved by this 'look-see' diagram:

The figure shows, as it were, a continuous rook's tour of the board. Removing any pair of opposite coloured squares will split the tour into two portions, each containing an even number of squares, alternately black and white, which can be covered by dominoes.

[Black, 1952, p. 157]

346. On average half the women will bear a girl first, and half will bear a boy, so on first births, the numbers of boys and girls will be equal. The women who bore a girl will then have no more children, while the half who bore a boy will continue to bear, having as second children, half boys and half girls, so that the balance of boys and girls is preserved.

Among the third children to be born within families, there will also be a balance of boys and girls, and so on.

The fact is that the number of families which consist of only one girl, which amounts to no less than one half of all families, will exactly balance the much smaller number of families containing several boys followed by a girl. Indeed, this amounts to no more than the fact that $\frac{1}{2} = \frac{1}{4} + \frac{1}{8} + \frac{1}{16} + \dots$

This solution assumes that the ratio of births of boys to girls is indeed one to one; the ratio actually favours boys very slightly, but this ratio in itself will never produce the surplus that the King requires, and his ingenious scheme will be of no help at all.

[Gamow and Stern, 1958, p. 20, communicated by Victor Ambartsumian, the eminent physicist]

347. ' "Very simple, my dear Watson," the Sultan chuckled. "As a matter of fact I expected this good news exactly on that day. My people, as I suggested before, may be too lazy to organize the shadowing of their wives for the purpose of establishing their faithfulness or unfaithfulness, but they have certainly shown themselves intelligent enough to resolve the case by purely logical analysis."

' "I do not understand you, Great Sultan," said the vizier.

' "Well, assume that there were not forty unfaithful wives, but only one. In this case, everybody with the exception of her husband knew the fact. Her husband, however, believing in the faithfulness of his wife, and knowing no other case of unfaithfulness (about which he would undoubtedly have heard) was under the impression that all wives in the city, including his own, were faithful. If he read the proclamation which stated that there are unfaithful wives in the city, he would realize it could mean only his own wife. Thus he would kill her the very first night. Do you follow me?"

' "I do," said the vizier.

' "Now let us assume," continued the Sultan, "that there were two deceived husbands; let us call them Abdula and Hadjibaba. Abdula knew all the time that Hadjibaba's wife was deceiving him, and Hadjibaba knew the same about Abdula's wife. But each thought his own wife was faithful.

' "On the day that the proclamation was published, Abdula said to himself, 'Aha, tonight Hadjibaba will kill his wife.' On the other hand, Hadjibaba thought the same about Abdula. However, the fact that next morning both wives were still alive proved to both Abdula and Hadjibaba that they were wrong in believing in the faithfulness of their wives. Thus during the second night two daggers would have found their target, and two women would have been dead."

' "I follow you so far," said the vizier, "but how about the case of three or more unfaithful wives?"

' "Well, from now on we have what is called mathematical induction. I have just proved to you that, if there were only two unfaithful wives in the city, the husbands would have killed them on the second night, by force of purely logical deduction. Now suppose that there were three wives, Abdula's, Hadjibaba's, and Faruk's, who were unfaithful. Faruk knows, of course, that Abdula's and Hadjibaba's wives are deceiving them, and so he expects that these two characters will murder their wives on the second night. But they don't. Why? Of course because his, Faruk's, wife is unfaithful, too! And so in goes the dagger, or the three daggers, as a matter of fact."

' "O Great Sultan," exclaimed the vizier, "you have certainly opened my eyes on that problem. Of course, if there were four unfaithful wives, each of the four wronged husbands would reduce the case to that of three and not kill his wife until the fourth day. And so on, and so on, up to forty wives."

' "I am glad," said the Sultan, "that you finally understand the situation. It is nice to have a vizier whose intelligence is so much inferior to that of the average citizen. But what if I tell you that the reported number of unfaithful wives was actually forty-one?" '

[Gamow and Stern, 1958, pp. 21–3]

348. The path of a point on the flange of a wheel moving on a rail looks like this:

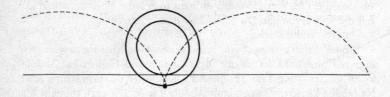

Round about the bottom of each loop, the point is moving in the opposite direction to its general motion. Therefore at any one moment, portions of the bottom of every wheel on the train are moving, albeit temporarily, back to Bristol.

It is also possible, but less certain, that portions of a rapidly moving wheel within the driving motor could be moving back to Bristol, without even being flanged.

349. The bicycle moves backwards. However, the pedal moves forwards, relative to the bicycle, so that the pedals as a pair are rotating, as would be expected, in the opposite direction to that required to move the cycle forwards.

350. This is an example of the 'pigeon-hole' principle. Consider one million boxes, numbered consecutively from 0, for the completely bald, to 999,999 for those people (who might just exist according to the information given) who have that many hairs on their head.

Place one slip for each person in the United Kingdom into the box corresponding to his or her number of hairs. Then at least one box must contain fifty slips of paper, corresponding to at least fifty people with that same number of hairs on their head.

351. (a) Yes, you have, if a child skipping can be described as running.
[Wells, 1983–6, Series 1, problem 97]

(b) The maximum overlaps in each case have the same area, so three-quarters of the triangle equals half the square, and the area of the triangle is 24 square inches.
[Wells, 1979, problem 57, part 3]

(c) When passing on a spiral staircase, where the insides are narrower and more difficult to walk on.
[Wells, 1979, problem 58, part 2]

(d) Tom thinks that Fred will not mind being five pence short, so presumably Tom will not mind being five pence short himself, which is what he would be if he gave Fred ten pence – which will naturally also satisfy Fred.
[Wells, 1979, problem 57, part 1]

352. The shortest route is shown overleaf. Most attempts to get through the maze end up with Theseus unable to leave without missing his last chance to turn left or right. His wander round the lower right corner is necessary to change the parity with which he approaches the north exit.

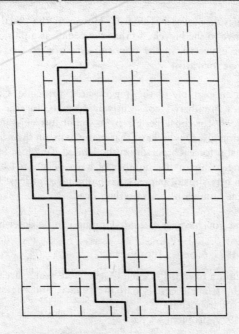

[Wells, 1979, problem 54]

353. I had dashed on to the platform at the rear end of the first train; I knew that the exit at my destination station was at the front end of the train, so I walked down the platform and reached the other end just as the next train arrived, and I got into the front carriage. Result: I had quite simply walked the length of the train at the first station, instead of at my destination station, and no time was wasted at all.

[Wells, 1979, problem 50]

354. Four men and four women were present. If you mark points for individual men and women, joining dancing partners by lines, then you will construct a skeleton of points and lines in which every region has four edges and three edges meet at every point. All skeletons with these properties are equivalent to either a single cube or to several entirely separate cubes. Since the soirée is described as intimate, it is reasonable to suppose that the skeleton is one cube only. There is one person for each of its eight vertices.

[Wells, 1979, problem 42]

355. This is the shape her husband produced. The size and proportions of the base are irrelevant as long as the base and sides of the box are rectangles. It will be possible to choose the height of the cut at each corner, and then make the cut with just one stroke of the sword, so that by using the appropriate corner, either one, two, three or four measures can be poured out. The technique is to fill the box so that the rice has a level surface, just comes up to the lip of the chosen corner, and leaves the opposite corner of the base rectangle just showing.

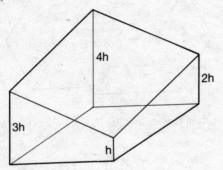

[Wells, 1979, problem 32]

356. Two pieces, cut as shown. These two squares can be interlaced to form the *mon*.

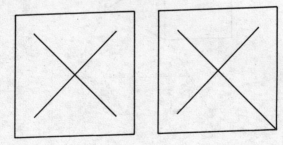

[Wells, 1979, problem 16]

357. This is an ordinary overhand knot, composed of just twenty-four cubes.

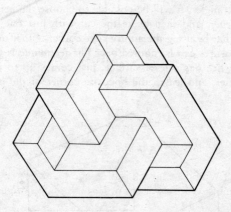

[Doug Engel, in *Games and Puzzles*, No. 36, May 1975]

358.

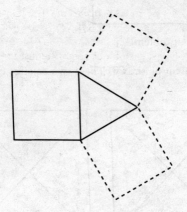

359. Three, one of the squares being the hollow square in the middle.

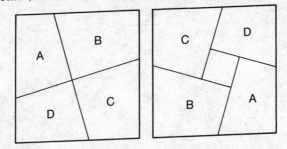

[After Adams, 1939, problem C12]

360. He lost. Winning multiplies his stake by $1\frac{1}{2}$; losing halves it. So three wins and three losses in any order multiply his original stake by

$$\left(\frac{3}{2}\right)^3 \left(\frac{1}{2}\right)^3 = \frac{27}{64}$$

So he loses $\frac{37}{64}$ of his £1, or 58 pence to the nearest penny.
[After Adams, 1939, problem B5]

361. Ninety-nine blocks, each block requiring just ten chops.
[Adams, 1939, problem C71]

362. The Red Lion.
[Adams, 1939, problem B32]

363. It can be done in fourteen moves: (1) C–a; (2) B–d; (3) E–b–B–e; (4) C–E–b–B; (5) A–c–C–a; (6) E–A; (7) C–e; (8) B–B–b–E; (9) C–B–d; (10) E–e–B–b; (11) C–B–e–A–c; (12) E–B–e; (13) D–b–B–d; (14) B–b.
[After Adams, 1939, problem C170]

364. Whatever the shape of the original triangle, the central triangle is $\frac{1}{7}$ of it, in area.

The simplest solution is based on the idea that it is possible to shear the original figure so that it becomes equilateral, without changing the relative sizes of its constituent triangles.

'Draw dotted lines as in the figure. We use the proposition that triangles with the same altitude are in proportion to their bases. Call each of the three smallest triangles the unit of area. Then each of the

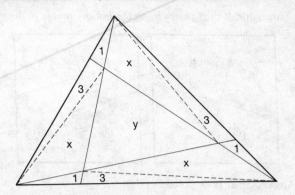

triangles marked 3 has three times the unit area. Of the four remaining triangles, mark the central one y and the others x. Then by the above proposition, $x + y + 3 = 3 (x + 1)$ and $2x + y + 7 = 3 (x + 5)$, when $x = 8$, $y = 16$. Since the whole triangle $= 52$, the central one $= 4/13$ of the original. (By comparing triangles, we also find that each line from a vertex to the division point of the opposite side, is divided in the ratio $4:8:1$.) By substituting n for 4 in this solution, we find similarly that the central triangle $= (n - 2)^2/(n^2 - n + 1)$ of the original (and each internal line from a vertex is divided into the ratio $n:n(n - 2):1$).'

[Graham, 1963, problem 52]

365. $69^2 = 4761$ and $69^3 = 328509$.

366. 'The dispute can be settled in the following manner: We give the priority of choosing the piece of ham to the third co-owner. She will choose, of course, the piece which according to her home balance is not less than either of the remaining two pieces. That is the piece whose value, according to her opinion, is not less than $4.00. Such a piece must exist because, by division of the whole into 3 parts, one of the parts cannot be less than $\frac{1}{3}$ of the total weight.

'Afterwards the second woman chooses her piece. She must also be satisfied because, after the third woman took her share, there remained at least one piece which, according to the balance in the shop on the corner, corresponded to a value not less than $\frac{1}{3}$.

'The first woman, who receives the remaining piece, must be satisfied, since she considered all the pieces to be of equal weight.'

[Steinhaus, 1963, problem 49]

367. The same. It is perfectly possible for the shortest giant and the tallest midget to be one and the same person. Indeed, if the men formed an array of k columns and m rows, then any man who has at least $k - 1$ colleagues shorter than himself and $m - 1$ taller than himself, can be in that double position, for a suitable arrangement of the men.

[After Steinhaus, 1963, problem 59]

368. Five. Suppose that it were possible for one town to be connected to six other towns. It is connected to one other town which is the closest town to itself. The other five connections are all to towns for which it is the closest other town. But if six points are arranged round a central point, the central point can only be *equal* closest to the other points, when they are arranged at the centre and vertices of a regular hexagon. As soon as the distances are adjusted, even very slightly, to make them all different, one of the ring of points will be nearer to another point than it is to the centre of the hexagon, contradicting the information given.

[After Steinhaus, 1963, problem 71]

369. Five is sufficient for a gallery of twenty walls at right-angles. The diagram shows a worst-case plan, which is nevertheless a common gallery design. One guard is needed for each of the rooms off the main corridor, and these guards can be placed so that the walls of the corridor also are always surveyed.

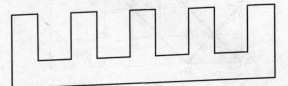

370. Very easily! Here are some especially composed sample figures to illustrate what can happen, adapted from Chapter 4 of Hugh ApSimon's *Mathematical Byeways in Ayling, Beeling and Ceiling* where there is a full discussion of the 'paradox'.

		Paul	Frank
First half of year	Runs	252	84
	Times out	4	1
	Average	63	84

Second half of year	Runs	84	252
	Times out	3	7
	Average	28	36
Whole year	Runs	336	336
	Times out	7	8
	Average	48	42

Note the uneven distribution of Frank's scores and the fact that if his scores for the first and second halves of the year had been switched, then Paul would have led in the first half of the year, Frank in the second, and the result would not seem at all paradoxical.
[ApSimon, 1984, p. 23]

371. It will definitely end up in one of the other three pockets. It is not possible for it to bounce for ever or end up in the pocket it started from.

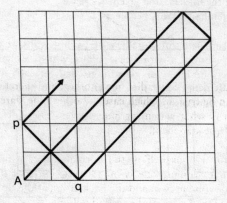

Suppose that it starts from pocket A, that the sides are integers p and q, as marked, and that 1 unit of distance is the length of the diagonal of one square.

Then, in travelling back and forth between the left-hand vertical edge and the right-hand vertical edge, the ball travels q, $2q$, $3q$ ... units, while in travelling back and forth between the top and bottom edges it travels p, $2p$, $3p$ ... units.

Therefore when a multiple of p first equals a multiple of q the ball will have travelled a whole number of times between the vertical edges and a whole number of times between the top and bottom edges – which is another way of saying that it will be in one of the corners.

Assuming that it stays in the corner, rather than bouncing back out, it can never retrace its path and end up in the corner from which it started.

[Mauldin, 1981, problem 147]

372. It must be struck parallel to one of the diagonals, and its total path back to its starting point is double the length of a diagonal.

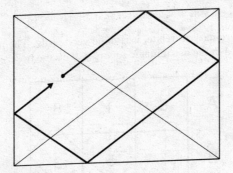

373. All the statements contradict one another, and therefore at most one of them can be true, in which case the other nine statements will be false, which is what statement nine asserts. Therefore statement nine is the only true statement.

374. Standing on a bookshelf in the normal order, which we can reasonably assume, volume one is on the left, its front flat against the back cover of volume two, and the back cover of the third volume is flat against the front cover of the second volume.

Therefore the bookworm only actually bores through the complete second volume, a distance of 8 cm.

375. If $a–bc$ means move coin a to touch coins b and c, then five moves are required from H to O: 1–56, 3–14, 4–58, 5–23, 2–54. But no less than seven moves are required to get back from O to H: D–CE, G–CD, D–CG, G–BD, C–AG, A–BE, E–FH.

[Brooke, 1963, No. 40]

376. Only three coins need be moved.

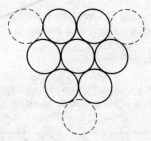

377. Extraordinary to relate, it is done very easily. The numbers 1 to 9 as arranged in the usual magic square will do the trick.

A	6	1	8
B	7	5	3
C	2	9	4

If Alan's dice have the numbers in the first row, with opposite faces of the dice showing the same number, and if Barry takes the numbers in row B, and Chris those in row C, then A beats C, beats B, beats A, each with the odds of 5 to 4. They could just as well take the columns instead.

[Berlekamp, Conway and Guy, 1982, p. 778]

378. The distance is equal to OX. This is obvious when the mirrors are at 45° and the incoming ray is instantly reflected at right-angles to the lower mirror, but the same result is true whatever the angle between the mirror, provided that that angle is an integral fraction of 45°. If it is not, then the ray cannot emerge along its initial path.

[Mackie, A. G. and Jellis, G. P., in *The Games and Puzzles* Journal, No. 4, March–April 1988, p. 58]

379. The balloon swings left with the car. This is most easily understood by considering that it is the air within the car that is heavy, and the coal-gas in the balloon relatively light, and naturally it is the heavier air that swings outwards, forcing the balloon inwards.

[Morris, 1970, p. 203, communicated by Gerald Stonehill]

380. The 20-litre barrel contains beer. The total quantity of wine is divisible by 3, since it was bought in two parts, one double the other. The total of the tens digits in the quantities is a multiple of 3, but the sum of the unit digits is 29. Therefore, the barrel containing the beer contains a multiple of 3, with 2 remainder. The 20 litre barrel is the only possibility.

The first customer bought the 15-litre and 18-litre barrels, and the second took the 16-litre, 19-litre and 31-litre barrels.

381. 'Twice four and twenty' could be either 28 or 48, of which only the first number is divisible by 7. Therefore Jill shot seven birds, and it was these seven who remained, as the others flew away.

[Morris, 1972, problem 39]

382. The buses do indeed run very regularly, and at equal intervals, every bus in one direction arriving, say, 1 minute after the previous bus in the opposite direction, and many minutes before the next bus in that same direction.

383. 16/64, 26/65, 19/95 and 49/98. There are many other such cancellations with larger numbers, such as 143185/17018560 = 1435/170560 and 4251935345/91819355185 = 425345/9185185.

[Domoryad, 1963, p. 35]

384. You need turn over only two cards. The first card shows a vowel, and so you must test whether it has an even number on the reverse. The third card shows an odd number, and this would contradict my claim if the letter on the other side were a vowel, so you must test card three also. Neither of the other cards can affect my claim either way.

385. He makes eight cigarettes and smokes them, leaving eight ends from which he makes two more cigarettes, a total of ten.

386. 5 pence and 10 pence. One of them is not a 10 pence piece, but the other is!

387. If the level of beer in the can is above the centre of gravity, the centre of gravity can be lowered by drinking more beer. Likewise, if the level of the beer is below the centre of gravity then the centre of gravity could be lowered by replacing some of the beer, to increase the amount of beer below the centre of gravity line.

Therefore, the centre of gravity will be a minimum when neither of the above cases applies, that is, when the surface of the beer and the centre of gravity coincide.

The exact position of this can only be calculated when more data are given about the size and weight of the can, and the density of the beer.

388. Two, one on either side.

389. After twenty-nine days.

390. Every time a pair of delegates shake hands, the total number of handshakes made increases by two. In other words, the total is always an even number. If an odd number of delegates shook hands an odd number of times, then however many shook hands an even number of times, the grand total would be odd, which is impossible. Therefore the conclusion of the puzzle follows.

391. No. The second man has not been placed anywhere.

392. 'Let $A = \sqrt{2}^{\sqrt{2}}$. If A is rational, it is the desired example. On the other hand, if A is irrational, then $A^{\sqrt{2}} = 2$ is the desired example.'

[*Litton's Problematical Recreations*, 1967, No. 9, problem 4]

393. It is only necessary for the lines to cross on the shoe. This is a simple and symmetrical solution.

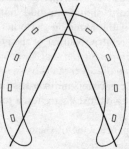

394. The woman was buying individual house numbers from an iron-mongers.

395. 'Since motion is relative, consider the hoop as fixed and the poor girl whirling around. The original point of contact on the girl traverses the diameter of the hoop twice, and this is the required distance.'

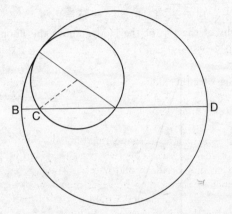

In the diagram, C is the original point of contact. As the inner circle rolls against the outer circle, C moves along the diameter BD.

[Trigg, 1985, problem 62]

396. John is a young lad and too short to reach the lift button for any storey higher than the sixth.

397. Call the slices A, B and C. Fry A and B on one side each, and then swop B for C and fry A on the other side and C on its first side. Now swap A for B and fry the second sides of B and C at the same time. Total time, 60 seconds.

398. The number of moves required is always $n - 1$, and cannot be shortened by, for example, fitting together pieces in chunks, and then joining the chunks together – though of course there are other good reasons for tackling a jigsaw puzzle in that manner.

[Trigg, 1985, problem Q29]

399. By similar triangles $\frac{x}{1} = \frac{1}{y}$, so $xy = 1$

Also $(x + 1)^2 + (y + 1)^2 = 4^2 = 16$

So $(x + y)^2 + 2(x + y) = 16$

and $x + y = -1 \pm \sqrt{17}$

Since $x + y$ is positive, $x + y = \sqrt{17} - 1$

and $x - y = \sqrt{(x + y)^2 - 4xy} = \sqrt{14 - 2\sqrt{17}}$

So $x = \frac{1}{2}(\sqrt{14 - 2\sqrt{17}} + \sqrt{17} - 1) = 2.76$

and the height of the top of the ladder above the floor $(x + 1)$ is approximately 3.76 m.

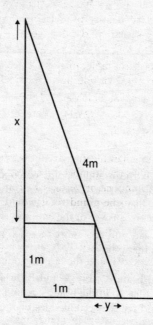

400. The following solution was contributed by a reader to L. A. Graham's *The Surprise Attack in Mathematical Problems*. It avoids the difficulties to which this deceptively simple-looking puzzle (like problem 399) so easily leads.

'One of our readers rose to this task as follows: "From the similar triangles $z = 8x/c$ and $y = 8x/d$, where c and d are the left and right segments of the required length x. Adding, we get $z + y = 8x^2/cd$ and multiplying, $zy = 64x^2/cd$ or $8(z + y)$.

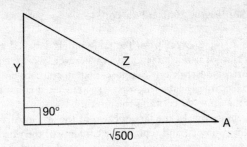

From the diagram, $900 - z^2 = 400 - y^2$, from which $z^2 - y^2 = 500$. We represent this last identity by a triangle, as in this figure, from which it is readily seen that $\cos A + \cot A = 500(z + y)/zy$, which from the previously derived relation equals $\sqrt{500}/8$ or 2.7951. We now merely look up a trig table, or use a calculator, to find the angle whose cosine and cotangent add up to the figure, and get $A = 27° 38' 30''$ approximately, from which $z = \sqrt{500}/\cos A = 25.24$ and the required value of $x = \sqrt{900 - (25.24)^2} = 16.2$ feet." '
   [Graham, 1968, problem 6]

401. The total number of edges meeting at all the vertices, if summed vertex by vertex, is double the number of edges, and therefore an even number. If an odd number of vertices had an odd number of edges meeting at them, then the grand total would also be odd – a contradiction. Therefore the answer to the question is 'no'.

402. If the street contains more than 100 houses but less than 1000, then Mr Jones lives at No. 204, and the street is numbered from 1 to 288.
   [After Beiler, 1964, p. 297, problem 23]

403. Thirty-six matches, because each match played eliminates one player, and thirty-six players must be removed to leave one winner. There are many ways to arrange the pairings, but the total number of matches played is not affected.

404. He kept ducks.

405. A man was sitting on a three-legged stool eating a leg of ham, when along came a dog and snatched the ham away. The man threw the stool at the dog, and recovered his ham.

406. The bottle costs 20½p and the cork ½p.

407. Take a single sweet from the jar wrongly labelled MIXED. You know that this jar is *not* mixed and so whatever sweet it contains tells you its correct description. Suppose that it contains aniseed balls. Then the jar wrongly labelled ANISEED BALLS must contain chocolate drops and the other jar contains the mixture.

Finally, empty one of the jars, fill two of the jars in succession with their correct contents, and replace the contents of the third jar, which is almost certainly much easier than switching the labels round.

408. There is no knot. Try it and see.
    [Budworth, 1983, p. 144]

409. 'We can build concentric hexagons containing 1, 6, 12, 18, 24, 30, 36, and 42 circles.

'When *R/r* (the ratio of the radius of the table to the radius of each circle) becomes sufficiently large there will be room for extra circles

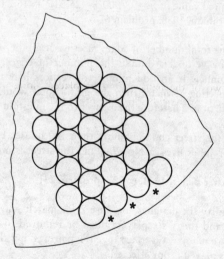

as indicated by * above. If there is an even number of circles per side in the last hexagon, an 'outsider' can be placed centrally if

$$R/r \geqslant \frac{1 + \dfrac{\sqrt{3}}{2}}{1 - \dfrac{\sqrt{3}}{2}} \text{ i.e. if } R/r \geqslant 13 \cdot 9$$

Two more "outsiders" can be put each side of this one if

$$\left[(R + r)^2 \left(\frac{\sqrt{3}}{2} \right)^2 + (2r)^2 \right]^{\frac{1}{2}} + r \leqslant R$$

i.e. if $0 \leqslant \dfrac{R^2}{r^2} - 14 \dfrac{R}{r} - 15$

i.e. if $0 \leqslant \left(\dfrac{R}{r} + 1 \right) \left(\dfrac{R}{r} - 15 \right)$

i.e. if $R/r \geqslant 15$

Hence in the example given three "outsiders" can be accommodated.
    'The number of saucers that can be placed on the table is:

$$1 + 6 + 12 + 18 + 24 + 30 + 36 + 42 + (3 \times 6) = 187'$$

[Kendall and Thomas, 1962, problem C12]

410. Two of the balls should be placed along one diagonal of the lower part of the cube, and the other two along the other diagonal in

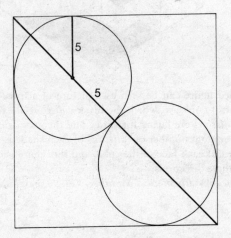

the upper portion, so that the balls are at the vertices of a regular tetrahedron.

Looking down on the cube at the balls in the lower layer, the length of the diagonal of the square is $2(5 + 5\sqrt{2})$, so the length of one side of the cube is $\sqrt{2}(5 + 5\sqrt{2})$ or $10 + 5\sqrt{2}$, approximately 17.07 inches.

By symmetry, the remaining two balls will fit as described in the top layer, just touching these two balls.

411. This beautiful argument is due to Guy David and Carlos Tomei.

Here is a full box. The *calissons* pointing in the three directions have been shaded grey or black, or left white.

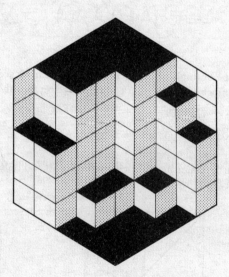

The shaded figure can be seen as a picture of a three-dimensional arrangement of cubes, in which the black *calissons* are the top faces, the white *calissons* are facing forwards, and the grey *calissons* face to the right. The total number of *calissons* of each shade is just equal to the total area of each face of the cube, and this figure is of course the same in all three directions.

[*American Mathematical Monthly*, Vol. 96, No. 5, May 1989]

412. He wraps the gun in canvas to make a rectangular package 2 metres long by at least 1.14 metres wide, so that the gun lies diagonally across the package. It is now an acceptable packet.

[Kraitchik, 1955, p. 40, problem 50]

413. Sixteen extra cigarettes can be accommodated:

'Twenty cigarettes are placed in the bottom layer. In the second layer instead of having twenty, we place nineteen, arranged as shown in the diagram. Then we continue with alternate layers of twenty and nineteen.

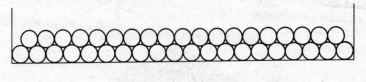

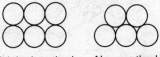

Original method   New method

'Suppose the diameter of a cigarette is 2 units. The second and subsequent layers, using our new method, will add only 1.732 units to the height. The depth of the box is 16 units, since it originally contained eight layers. With our new method we shall get nine layers: $2 + 8 \times 1.732$ is equal to 15.856. So with five layers of twenty and four layers of nineteen we shall get 176 cigarettes into the box.'

[Brandreth, 1984, p. 55]

414. Move A a little way away, and press down on coin B to stop it moving. Flick A against B and C will move away, allowing A to be placed between B and C.

415. 601 is less than 2^{10} but more than 2^9, so they played ten games. The gross total of stakes was $1 + 2 + 4 + 8 + \ldots + 2^9 = 1023$ cents. If the gross winnings were Steve x cents and Mike y cents, then $x + y = 1023$ and $x - y = 601$; solving, $x = 812$ and $y = 211$.

Only the stakes in the first, second, fifth, seventh, and eighth games add up to 211 cents. So those are the games that Mike won.

[Madachy, 1966, problem 7, p. 177]

416. 'Solving the inequalities simultaneously, we find that $56\frac{1}{4}$ $> N > 53\frac{1}{3}$. Knowing that $N = 54$, 55 or 56, the other inequalities lead to the unique solution: $P = 26, E = 19, N = 55$.'

[*Litton's Problematical Recreations*, 1967, No. 11, problem 10]

417. From the numbers given it is clear that the number chosen cannot start with a 0. The number of allowable pandigital numbers is therefore the total number, including initial zero, less the number starting with a zero: it is $10! - 9! = 9 \times 9!$

There are 9,000,000,000 numbers in the range, so the probability is $(9 \times 9!)/(9 \times 10^9) = 362,880/1,000,000,000 = 0.00036288$, or less than 1 in 2500.

418. One quarter. This is proved by an elegant geometrical argument.

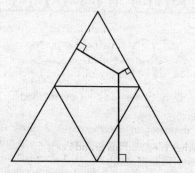

Consider an equilateral triangle, whose altitude is equal to the length of the stick. For any point inside the triangle, the sum of the perpendiculars from the point to the sides of the triangle is a constant, equal to the altitude of the triangle, that is, to the length of the stick. (This is also true for points outside the triangle, provided the appropriate perpendiculars are counted as negative in length.)

If the point lies inside any of the shaded triangles, then one perpendicular will be longer than the sum of the other two, and no triangle is possible. Therefore, the divisions of the stick for which a triangle can be formed correspond to the points in the central triangle, which is one quarter of the whole.

419. One quarter of the length. The breaking point of the stick, since the break is made at random, is equally likely to occur anywhere along the stick, and so the shorter of the two pieces is equally likely to be any length from zero to one half, and will on average be a quarter.

(The stick can hardly be said to be broken if the break is at one end, but the probability that it will be broken exactly at an end is zero, and equal to the probability that it will be broken *exactly* at the half way point.)

[Mosteller, 1987, problem 42]

420. There are five equal gaps between six strikes, so the time between strikes is 3 seconds. When striking 12 there will be $3 \times 11 = 33$ seconds between the first and last strikes.

421. Let the distances of the four towns along the road, from some fixed point, be a, b, c and d, in the order in which they are named. Then the first instruction takes the treasure seeker to $\frac{1}{2}(a + b)$ and the second to $\frac{1}{2}(a + b) + \frac{1}{3}(c - \frac{1}{2}(a + b)) = \frac{1}{3}(a + b + c)$.

By a similar calculation, the fourth instruction takes the seeker to $\frac{1}{4}(a + b + c + d)$, and this symmetrical expression can be calculated without knowing which town is which.

In fact, it is the centre of gravity of the four points, and from the distances given, is located at X, which is 1 mile from B and 7 miles from C.

[NCTM, 1965, problem 91]

422. The grandfather is 66 and his grandson is 6.
[NCTM, 1965, problem 120]

423. The grandson was born in the twentieth century, in 1916, and was 16 years old in 1932. The grandfather was born in the nineteenth century, in 1866, and was 66 in 1932.
[Perelman, 1979, problem 5]

424. A third diagonal can be added, to complete an equilateral triangle, any one of whose angles is 60°.

425. The circles are equally spaced and their areas are 1, 4, 9, 16, 25, 36 and 49. Since $9 = 25 - 16$, and $25 - 1 = 49 - 25$, the inside of circle C is equal in area to the annulus between circles D and E, and the region between circles A and E equals in area the annulus between E and G.

426. 'Fred's bets were in the proportion $1/(3 + 1)$, $1/(4 + 1)$, $1/(7 + 1)$, $1/(9 + 1)$ and $1/(39 + 1)$, the last being placed on each of five horses. These fractions add up to 0.8, and no matter which horse wins, the winnings plus the returned winning stake total 1.00, i.e. the profit is 0.2. Hence Fred's total stake was £800 and his profit £200.

'It's no use trying it yourself – real bookies fix the odds better than that!'

Fred's stakes were: £250 on Bonnie Lass, £200 on Golden Stirrup, £125 on Two's a Crowd, £100 on Greek Hero, and £25 on each of the five others.

[Eastaway, 1982, p. 94]

427. Place the set square so the vertex at the right-angle lies on the circle.

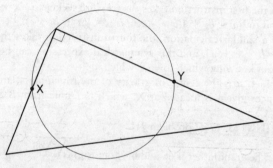

Because of the property that 'the angle in a semi-circle is a right-angle', the line XY will be a diameter of the circle. By repeating the process a second diameter is obtained, and the centre is their intersection.

428. 'Let A be the angle in minutes between the hands, with the minute hand ahead of the hour hand. If the hour hand moves through two-thirds of angle A, the minute hand must move through $(60 - A) + \frac{1}{3}A$. Since the minute hand moves twelve times as fast as the hour hand, $12(\frac{2}{3}A) = 60 - \frac{2}{3}A$ and $A = 6\frac{12}{13}$ minutes. The elapsed time is $12 \times \frac{2}{3} \times 6\frac{12}{13} = 55\frac{5}{13}$ minutes.

'It is very interesting that if the hour hand only moves through one-third of the angle so that the minute hand ends up ahead of the hour hand, an identical result is obtained. While the hour hand moves through one-third of the angle B, and the minute hand moves through $(60 - B) + \frac{2}{3}B$. Proceeding as before, $12(\frac{1}{3}B) = 60 - \frac{1}{3}B$ and $B = 13\frac{11}{13}$

minutes. The elapsed time is now $12 \times \frac{1}{3} \times 13\frac{11}{13} = 55\frac{5}{13}$ minutes.

'In order to rule the lines as soon as possible after 3 o'clock, the angle chosen must be the smaller one, $6\frac{12}{13}$ minutes. Then the hour hand movement X must be one-twelfth the movement of the minute hand: $15 + 6\frac{12}{13} \times X$; $12X = 21\frac{12}{13} + X$, and $X = 1\frac{142}{143}$ minutes. The minimum time to reach this configuration is $12 \times 1\frac{142}{143} = 23\frac{131}{143}$ minutes.'

[Contributed by Marlow Sholander to Graham, 1968, problem 17]

429. Nine sticks are sufficient, the same number required to make a pentagon.

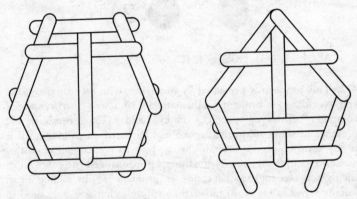

[NCTM, 1978, p. 184]

430. Group the pennies in units of four, touching at the ends. To allow the end pennies to touch three pennies, the ring must be completed, which can be done with five groups, a total of twenty pennies.

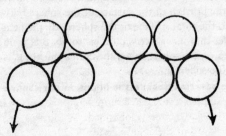

[Kendall and Thomas, 1962, problem A18]

431. 'Put your first and second fingers on 1 and 2, bring them round to the corresponding position on the right-hand side. Then push the six coins bodily to the left, leaving coins 1 and 2 in the position shown.'

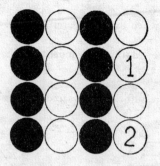

[Abraham, 1933, problem 124]

432. This problem is presented by Mosteller, who records that John von Neumann, a brilliant mathematician who was also a walking computer, solved the problem 'in his head in 20 seconds in the presence of some unfortunates who had laboured much longer'. If it is any consolation – it probably won't be – to readers who find this anecdote depressing, von Neumann also once solved the problem of the fly between approaching trains (cf. problem 288) by adding up an infinite series in his head, rather than spotting that you only need to divide the distance between the trains by the fly's speed.

Mosteller points out that some simplifying assumptions are needed to solve this problem. It would be extremely complicated, and require empirical testing, if for example the elasticity of the table were taken into account. He suggests inscribing the coin in a sphere, as in the figure, and assuming that the chance that the coin falls on its edge, is the ratio of the portion of the surface of the sphere between the edges of the coin to the total surface of the sphere. By the beautiful theorem of Archimedes that was inscribed on his tomb, this zone will have one third the area of the sphere when the coin's thickness is one third the diameter of the sphere.

By Pythagoras, this thickness is about 35 per cent of the diameter of the coin:

$$R^2 = r^2 + \frac{1}{9}R^2$$

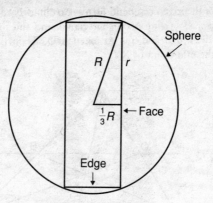

or

$$\frac{8}{9}R^2 = r^2$$

$$\frac{R^2}{9} = \frac{r^2}{8}$$

$$\frac{1}{3}R = \frac{\sqrt{2}}{4}r \approx 0.354\,r$$

[Mosteller, 1987, problem 38]

433. Take a rectangular sheet of paper and use the ruler to draw two parallel lines, as in the figure. Then placing the ruler as indicated, the angle BAC will be 30°.

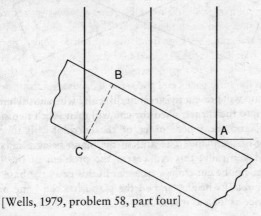

[Wells, 1979, problem 58, part four]

434. XY is the given segment. Draw two lines parallel to it, as in the figure. Take any point A on the second line and join it to X and Y, and then draw XB and YC to meet in O. Then the line AO passes throught the mid point of XY.

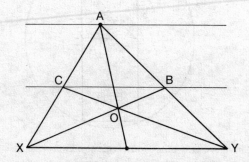

Another solution, which involves placing opposite sides of the ruler against opposite ends of the line segment, an operation sometimes considered to be illegal, is illustrated in the figure below. The dotted diagonal of the parallelogram bisects the original segment.

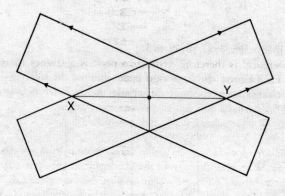

435. The piano will present the greatest difficulty when it is symmetrically wedged into the corner. Two corners will then touch two sides of the corridor, and the inner corner of the corridor will touch the opposite side of the piano. The area of the piano being a maximum when viewed vertically, this reduces to the problem of the largest rectangle that can be cut from a triangle. In this case, the base of the triangle is double its height, and so the piano has the same proportions: it is twice as long as it is wide.

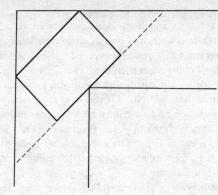

436. There are two other four-digit numbers with this property, 2025 and 9801, of which only 9801 has distinct digits.

437. 'Calling the unknown lengths of the scale-arms a and b, and the parcel's true weight x ounces, by the principle of the lever

$$bx = 28\tfrac{1}{4}a$$
$$ax = 36b$$
from which $\quad x^2 = 1017$
and $\quad x = 31.89\ldots$

The parcel is therefore under two pounds and goes for sixpence. If I had adopted the happy-go-lucky method of striking an average between the recorded weights, I should have arrived at the figure of $32\tfrac{1}{2}$ oz, a weight demanding a ninepenny stamp.'

[Williams and Savage, n.d., problem 57]

438. One weighing is enough. Take one coin from the first box, two from the second, three from the third, and so on, up to ten coins from the tenth box. Then the weight of all these coins will fall short of the total if they were all sound, by 2 gm times the number of the box containing the duds.

439. Three weighings are necessary. This is one schema:

In left-hand pan					*In right-hand pan*			
1	2	3	10		4	5	6	1
1	2	3	11		7	8	9	10
1	4	7	10		3	6	9	12

Denote a result in which the pans balance by 0, one in which the right-hand pan falls by r, and one in which the left hand pan falls by l . . . 'Interpretation of a weighings record is then a simple matter. If the result is 000, all coins must then have the same weight. If the result is $r0l$, we look for a coin with programme $r0l$, find this to be coin No. 4, and conclude that coin No. 4 is too heavy. However, if the result is $0lr$, for example, we find no coin with this programme, and conclude that it is the coin with the opposite programme $0rl$ – namely coin No. 7 – which then is too light. The extension to the analogous problem for 39 coins with four weighings, or for 120 coins with five weighings, and so on, is not difficult.

'This solution can be applied when thirteen coins are in question, if it is known that *exactly one* of them is false. The weighings result 000 then indicates that coin No. 13 is false – a unique case in which there is no indication whether the false coin is too heavy, or too light.'

[Sprague, 1963, problem 8]

440. Yes. Jack accepted Fred's bet and handed over two pounds. Fred welshed on his bet, and handed over one pound, making a profit of one pound.

441. This table shows the payoffs for Red, according to Black's response. The only bids for Red which prevent Black from winning are 5p and 6p, and of these 6p is the better: if Black foolishly bids high, then Red will make a profit of 1p with his 6p bid, but would break even by bidding 5p.

	BLACK BID					
	3p	4p	5p	6p	7p	8p
3p	0	-2	-2	-2	-2	-2
4p	2	0	-1	-1	-1	-1
5p	2	1	0	0	0	0
6p	2	1	0	0	1	1
7p	2	1	0	-1	0	2
8p	2	1	0	-1	-2	0

RED BID (labels at left of rows 5p/6p)

PAYOFF to RED (p)

The result holds true regardless of the number of players (assuming that the high bidders divide the pool and the liabilities equally). Always bid one unit more than the stake.

[Silverman, 1971, problem 42]

442. Of the eight sandwiches contributed, each man ate $8/3 = 2\frac{2}{3}$ sandwiches. Therefore Jones contributed $2\frac{1}{3}$ of his sandwiches to Watson's lunch, while Smith contributed only $\frac{1}{3}$ of a sandwich, only $\frac{1}{7}$ as much. So Smith was right to insist that Jones was wrong, and the £2 was actually divided into £1.75 for Jones and 25p for Smith.

443. Flowers cost 10p, whilst I.P.A. costs less than 10p. The first stranger put down a 10p piece on the counter. The second man put down a 5p piece, and five pence in copper coins.

[Kendall and Thomas, 1962, problem A9]

444. There are $8 \times 8 = 64$ individual squares, plus $7 \times 7 = 49$ squares composed of four individual squares, plus $6 \times 6 = 36$ squares of nine individual squares . . . and so on, making a grand total of:

$$8^2 + 7^2 + 6^2 + 5^2 + 4^2 + 3^2 + 2^2 + 1^2 = 204 \text{ squares.}$$

445. This is one of the original constructions described by Lorenzo Mascheroni in 1797 in his *Geometria del Compasso*, in which he demonstrated that you do not need a ruler or straight edge to do geometry: an ordinary pair of compasses will suffice by themselves. He demonstrated this surprising fact by showing how all the basic constructions of Euclidean geometry could be so performed, from which it follows that all the more complex constructions based on them are also solvable by compass only.

Draw two circles centred on the given points, A and B, and with radius AB. From B, without changing the compass setting, mark off

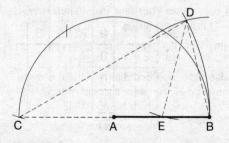

three arcs to find C, the opposite end of the diameter BAC. With radius CB and centre C, draw an arc to intersect the circle centred on B at D. Finally with radius AB and centre D, draw an arc which will intersect AB at E, the mid-point of AB.

The isosceles triangles DBE and CDB have the same base angle at B, and are therefore similar. Since DB is one half of BC, BE is one half of BD = BA.

[Graham, 1963, problem 11]

446. One move. Pick up the third tumbler from the left and pour its contents into the last tumbler.

[Always, 1965]

447. Yes, in six possible different ways. The eight arrangements of heads and tails can very conveniently be displayed at the corners of a cube, in such a way that each arrangement can be changed into an adjacent arrangement by moving to an adjacent vertex of the cube.

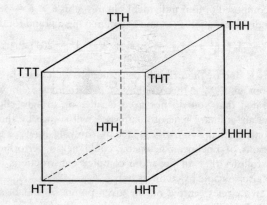

There are three ways to set out from HHH on your journey, and two choices of route after your first stop. Thereafter, the rest of the route is forced. This is one solution:

HHH   THH   THT   HHT   HTT   HTH   TTH   TTT

448. The two dates give 1 as first digit of 11 and 8 *across*.

15 (a cube) must be 27; for since 9 *down* ends in 1, 64 would give a 3 for end of 7 *d*. (a square), which is impossible.

∴ 16 *ac*. is 16 and 9 *d*. is 11.

Since 10 *d.* = 10 *ac.* × 11, 10 *ac.* must end in 2.

We can now calculate 14 *ac.* (perimeter of D.M.) to be 792.

Inspection of col. 2 shows that 12 *d.* = 19.

11 *ac.* must be 191* and 3 *d.* must begin with 1 or 2.

No. of roods being integral (see note), 1 *ac.* is an even multiple of 10, ∴ it ends 20.

Length and breadth of D.M. must be two numbers whose sum is 396 and product ends in 20, and as between them they contain the factor 11 twice over, they can only be 220, 176.

It is now possible to fill in the diagram thus far:

3	8	7	8	0	■	
	■			2	■	4	4
	■			■	3	5	2
■		1	6	1	0	■	■
		2	■	1	9	1	3
	■		■		7	9	2
2	7	■		1	6	■	5

'Shillings per acre' begins with 3, ∴ price per acre is between £15 and £20; and there are 8 acres.

∴ 4 *d.* is 142 and 1 *d.* is 355.

The 'down' number that = an 'across' number must be 10 *d.* And 11 times Farmer's age cannot = 352.

∴ it is 792.

Finally, we want a square for 2 *d.*, 7**6. This can only be 84² or 86². But 84² = 7056, which will not do for 5 *ac.*

∴ 2 *d.* = 7396

and Mrs Grooby's age is 86.

---

449. Every row and every column must contain at least one black square, so solutions with eighteen squares are easy and numerous. The pattern on the left exploits the knight's move, to produce a solution in seventeen squares, and no doubt for a sufficiently large initial board this solution or its reflection will be maximal. However,

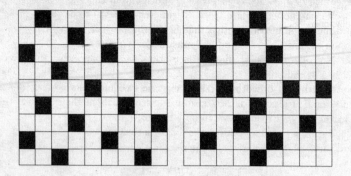

by taking advantage of the limited size of the given board, a solution in only sixteen squares is possible, as illustrated on the right.

[Wells, 1979, problem 63]

450. The solution is 242/303 = 0.7986798679867986 . . .

> Let F =   .TALKTALK . . .
> Then 10000F = TALK.TALK . . .   Subtracting,
>   9999F = TALK.

Then (EVE)/DID = F = (TALK)/9999. Therefore (TALK)/9999 when reduced to lowest terms equals (EVE)/DID, hence the denominator, 'DID' is a 3-digit factor of 9999, namely 101, 303, or 909.

(*a*) Assume DID = 101. Then (EVE)/101 = (TALK)/9999 or TALK = (EVE)99 = (EVE)(100 − 1) = EVE00 − EVE. This leaves an E as the first digit which therefore cannot be T. Therefore DID ≠ 101.

(*b*) Assume DID = 909. Then (EVE)/909 = (TALK)/9999 or TALK = (EVE)11 = (EVE)(10 + 1) = EVE0 + EVE and this time the units digit is E and therefore cannot be K. Therefore DID ≠ 909.

(*c*) Hence if there is a solution DID = 303. Since F is a proper fraction, EVE can be only (1*1) or (2*2), leaving for trial only 121; 141; 151; 161; 171; 181; 191; also 212; 242; 252; 262; 272; 282; 292. All these except 242 repeat, in the quotient, a digit that appears in the dividend so that finally 242/303 = .79867986 . . . is the only solution.

[Beiler, 1964, p. 301, problem 66]

451. The number is 102564: 102564 × 4 = 410256. The answer is easily found by starting with the fact that it ends in a 4:

4

and multiplying by 4, so that the next figure to the left is 6, with 1 to carry:

64

Multiplying the 6 by 4 and adding the carry, the next figure is 5, with 2 to carry:

564

and, since $(4 \times 5) + 2 = 22$, the next is 2 with 2 to carry:

2564

$(2 \times 4) + 2 = 10$, so place 0 and carry 1:

02564

$(0 \times 4) + 1 = 1$, so place the 1, with no carry:

102564

This is the solution.

452. The most likely total is 13. For every way in which you could end up with a total of 14 or more, there is a way of ending up with 13: all you have to do is to throw one less, and this is certainly possible because you cannot first exceed 12 and arrive at a total of 14 or more by throwing a 1.

Therefore the number of ways of reaching 13 is at least as great as the number of ways of reaching any higher total. But there are also ways to first exceed 12 and reach 13, for example by starting with 12 and throwing a 1, which do not have any matching throws for higher totals. Therefore there are more ways of getting to 13 and the probability is greatest that your total will be 13.

[Honsberger, 1978, p. 42]

453. About 73 per cent more. The figure shows the shape of the triangular cake and how it was cut. The original triangle and all its parts have angles of 120°, 30° and 30°. The ratio of a larger to a smaller piece is $\sqrt{3}$.

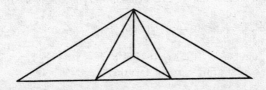

454. Three pieces are sufficient. Here are two solutions.

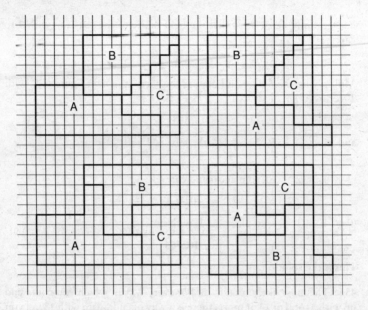

[Eastaway, 1982, p. 123]

455. Let the original cheque be for x dollars and y cents. Then

$$100y + x - 68 = 2(100x + y)$$

or
$$98y - 68 = 199x$$

If $98y - 68$ is a multiple of 199, so will be its double, $196y - 132$, which is $3y - 63$ short of $199y - 199$.

This difference will be zero if y is 21, and this is the smallest possible value for y. (It could also be $21 +$ any multiple of 199.) So the original cheque was for \$10.21.

[Beiler, 1964, p. 294, problem 10]

456. Ignoring reflections and rotations, this is the unique solution.

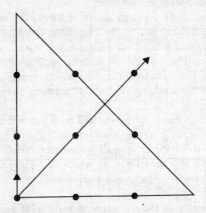

The original instructions did not forbid going outside the original square, so that is clearly allowed.

457. This is the unique and elegant solution, again ignoring rotation, which produces just one other solution.

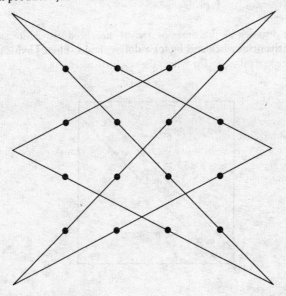

[Schuh, 1968, p. 340]

458. 'At nobody. Fire your pistol in the air, and you will have the best chance among all three truellists!

'Certainly you don't want to shoot at Black. If you are unlucky enough to hit him, Gray will polish you off on the next shot. Suppose you aim at Gray and hit him. Then Black will have first shot against you and his overall probability of winning the duel will be $\frac{6}{7}$, yours $\frac{1}{7}$. Not too good. (The reader is invited to confirm Black's winning probability of $\frac{6}{7}$ by summing the infinite geometric series: $\frac{2}{3} + (\frac{1}{3})(\frac{2}{3}) (\frac{2}{3}) + (\frac{1}{3})(\frac{2}{3})(\frac{1}{3})(\frac{2}{3})(\frac{2}{3}) + \ldots$)

'But if you deliberately miss, you will have first shot against either Black or Gray on the next round. With probability $\frac{2}{3}$, Black will hit Gray, and you will have an overall winning probability of $\frac{3}{7}$. With $\frac{1}{3}$ probability Black will miss Gray, in which case Gray will dispose of his stronger opponent, Black, and your overall chance against Gray will be $\frac{1}{3}$.

'Thus by shooting in the air, your probability of winning the truel is $\frac{25}{63}$ or about 40 per cent. Black's probability is $\frac{8}{21}$ or about 38 per cent. And poor Gray's winning probability is only $\frac{2}{9}$ (about 22 per cent).

'Is there a lesson in TRUEL which might have application in the field of international relations?'

[Silverman, 1971, problem 79]

459. Let x and y be the times of arrival measured in fractions of an hour between 5 o'clock and 6 o'clock. The shaded figure in this graph shows the arrival times for which the duellists meet.

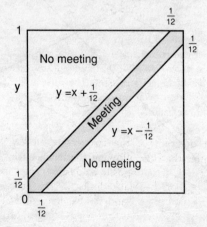

The area representing non-meeting is $(11/12)^2$ and so the area representing meeting is 23/144, and the chance that they actually do fight is a little less than 1/6.

[Mosteller, 1987, problem 26]

460. This is one solution. Numbering the matches 1–15 from left to right: move 5–1, 6–1, 9–3, 10–3, 8–14, 7–14, 4–2, 11–2, 13–15, 12–15. The diagram shows the position after the first five moves of the solution.

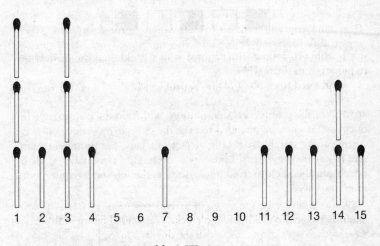

[Kordemsky, 1972, problem 57]

461. Three moves are sufficient. Numbering the cups 1–7, in any order you choose, you could invert 1, 2 and 3; 3, 4 and 5; 3, 6 and 7.

If however, the rules specified that four had to be inverted at each turn, then the problem is impossible, because after each inversion there will always be an even number of cups the right way up, and never an odd number, such as 7.

462. Here is a diagram of the twenty-five desks, shaded like a chessboard. Notice that by following the teacher's instructions each pupil moves to a desk of an opposite colour.

Peaky sat in one of the black desks, so on Tuesday there were twelve pupils in black desks and twelve in white, and they could follow the teacher's instructions, but when Peaky returned there were thirteen pupils in black desks and only twelve in white, and so it was

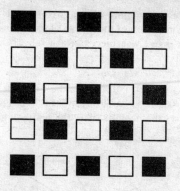

not possible to follow instructions, which Peaky had the misfortune to point out to the teacher.

[Adapted from NCTM, 1978, problem 115]

463. Since the problem rejects solutions which produce a square with thickness, it must be intended to take the cube as its surface only. By cutting along suitable edges the surface of a cube can be flattened into one piece, in fact into a hexomino, in eleven different ways, all of which can then be dissected into a square, though some require more than four pieces.

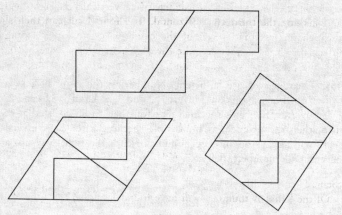

This 'Z' shape makes a cube and requires only four, by standard methods. The top figure is composed of six identical squares. The slanting line passes through its centre at approximately 54·7°. (Variation from this figure will produce a rectangle rather than a square.)

The second figure is composed of the two portions of the first figure rearranged. The line through the centre of this figure is perpendicular to the sloping edges. These pieces are then rearranged to form the third figure.

[*Games and Puzzles Journal*, No. 3, 1988, page 41]

464. With four vertical slices, and assuming that Jane's cake is not of some extraordinary shape, a maximum of eleven pieces is possible.

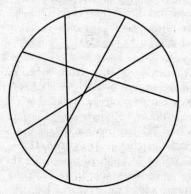

If the cuts need not be vertical, then the first three cuts can create eight pieces, for example by slicing twice vertically at right-angles, and making the third cut horizontal. The fourth cut can then slice through no less than seven of the eight pieces, making a total of fifteen pieces. For example, take the plane cut which passes through the mid-points of six edges and six eighth-cubes, and displace it slightly.

465. Sunday.

[Abraham, 1933, problem 10]

466. Suppose that there are E errors in total and that the first proof-reader, A, finds $\frac{1}{x}$ of the errors, and the second, B, finds $\frac{1}{y}$ of them. Then $\frac{E}{x} = 30$ and $\frac{E}{y} = 24$.

Of the $\frac{E}{x}$ that A found, B will have found $\frac{1}{y}$, and so from the errors found by both

$$\frac{E}{xy} = 20$$

So the total expected number of errors is $(30 \times 24)/20 = 36$.

Between them they found $30 + 24 - 20 = 34$, so it is expected that on average only two errors remain undetected by either of them.

467. Four. The sentence contains three spelling mistakes, plus the false claim that it only contains one mistake, making a total of four mistakes.

The second question, paradoxically, cannot be answered. It contains only two spelling mistakes but claims to contain three mistakes; therefore that claim is wrong and it actually contains three mistakes – except that if it contains three mistakes then the claim that it contains three mistakes is correct, and so it only contains the two spelling mistakes, in which case . . .!

468. 'Let n be the number of steps visible when the escalator is not moving, and let a unit of time be the time it takes Professor Slapenarski to walk down one step. If he walks down the down-moving escalator in 50 steps, the $n - 50$ steps have gone out of sight in 50 units of time. It takes him 125 steps to run up the same escalator, taking five steps to every one step before. In this trip, $125 - n$ steps have gone out of sight in 125/5, or 25, units of time. Since the escalator can be presumed to run at constant speed, we have the following linear equation that readily yields a value for n of 100 steps:

$$\frac{n - 50}{50} = \frac{125 - n}{25}$$

[Gardner, 1966, Chapter 14, problem 4]

469. The best location is at X, on a vertical line which has one kiosk to its left and one to the right, and on a horizontal line which has one kiosk above and one kiosk below. Moving the kiosk to Y, for example, would reduce the horizontal distance travelled by the owner of one kiosk while increasing the distance travelled by two of them.

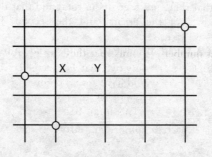

Of course, it cannot be denied that the criterion proposed, merely to minimize the total distance travelled, has turned out to be unfair on the owner of the north-east kiosk, who will travel much further than either of the others – but that injustice was no part of the puzzle.

[Sprague, 1963, problem 1]

470. The chance of receiving a second Nobel prize only depends on the total number of living Nobel-prize winners if you know for a fact that the Nobel prize committee have decided to honour, again, one of that select group. But Nobel prizes are awarded quite independently of past awards, and so the actual chance of being awarded two Nobel prizes is one chance in a billion times a billion (if you suppose, which is implausible, that every human being on earth has an equal chance of an award).

Linus Pauling's Nobel prizes, incidentally, were the prize for Chemistry and the Nobel Peace prize.

[Morris, 1970, problem 61. Morris does not give a satisfactory solution]

471. The fallacy is exposed when you use the argument of the problem to prove that any two horses are of the same colour. Removing each of the horses in turn only leaves the other one horse, and the set of $N - 1$ horses which are all of the same colour as each other, and the same colour as the horses removed, has one member.

If only it were possible to conclude that any pair of horses were the same colour, then it would indeed – and very obviously – follow that any three, four, five, etc., horses were of the same colour.

472. The surgeon was the boy's mother.

473. His son.

474. The points are: A – 8
B – 14
C – 9

The greatest number of points that they can get for wins or draws is: A – 5, B – 10, C – 5. ∴ only two matches were played (not one, for in that case one of them would have got *no* points).

Suppose that B drew two matches, against A and C, then B's total of goals should equal (A + C)'s total. But B's total is 4, and A's is 3 and C's is 4.

∴ A v. C was a draw and B won one.

Either A or C played only one game, and since A got 3 goals and C got 4, A v. C must have been 3–3. And C scored 1 goal against B. ∴ B v. C was 4–1.

Complete solution
A v. C   3–3
B v. C   4–1
    [Emmett, 1976, problem 82]

475. Supply four more identical triangles, and you will see that the marked dots are on the diagonal of the completed square:

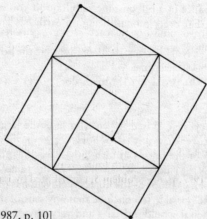

    [Wells, 1987, p. 10]

476. 9^{9^9} is by convention interpreted as $9^{(9^9)} = 9^{387,420,489}$.

This number, if written out without using index notation, contains more than 360 million digits. Nevertheless, it is possible to find its final two digits quite easily. The final digits of powers of 9 follow a cycle, which can be found by multiplying the last two digits alone, by 9, again and again:

Power of 9	Last digits
1	9
2	81
3	29
4	61
5	49
6	41

7	69
8	21
9	89
10	01
11	09
..	..

Very conveniently, 9^{10} ends in 01, and the cycle starts again. So $9^{387,420,489}$, which is 9^9 multiplied by a power of 9^{10}, ends in the same pair of digits as 9^9, that is, 89.

By a similar argument, based on the repeating sequence of the last three digits, the last three digits of the power could also be calculated.

477. 'Among the five married couples no one shook more than eight hands. Therefore if nine people each shake a different number of hands, the numbers must be 0, 1, 2, 3, 4, 5, 6, 7, and 8. The person who shook eight hands has to be married to whoever shook no hands (otherwise he could have shaken only seven hands). Similarly, the person who shook seven hands must be married to the person who shook only one hand (the hand of the person who shook hands only with the person who shook eight hands). The person who shook six must be married to the person who shook two, and the person who shook five must be married to the person who shook three. The only person left, who shook hands with four, is my wife.'

[Gardner, 1977, p. 69, problem 5, contributed by Lars Bertil Owe]

478. Between races, they agreed to swop horses, so that each was then mounted on the other's horse, and determined to leave his own horse trailing second.

479. 'Start the 7- and 11-minute hourglasses when the egg is dropped into the boiling water. When the sand stops running in the 7-glass, turn it over. When the sand stops running in the 11-glass, turn the 7-glass again. When the sand stops again in the 7-glass, 15 minutes will have elapsed.'

[Gardner, 1981, p. 190, problem 8.1, contributed by Karl Fulves]

480. This problem was first posed by T. P. Kirkman, a notable amateur mathematician, in 1847, and repeated in the *Lady's and Gentleman's Diary* for 1850. Only in 1969 did Ray-Chaudhuri and Wilson show that there is a solution in the general case, in which the number of girls is any odd multiple of 3.

Sun.	Mon.	Tues.	Wed.	Thurs.	Fri.	Sat.
0, 5, 10	0, 1, 4	1, 2, 5	4, 5, 8	2, 4, 10	4, 6, 12	10, 12, 3
1, 6, 11	2, 3, 6	3, 4, 7	6, 7, 10	3, 5, 11	5, 7, 13	11, 13, 4
2, 7, 12	7, 8, 11	8, 9, 12	11, 12, 0	6, 8, 14	8, 10, 1	14, 1, 7
3, 8, 13	9, 10, 13	10, 11, 14	13, 14, 2	7, 9, 0	9, 11, 2	0, 2, 8
4, 9, 14	12, 14, 5	13, 0, 6	1, 3, 9	12, 13, 1	14, 0, 3	15, 6, 9

This is one solution. Further discussion and references will be found in Rouse Ball, 1974, p. 287.

481. Professor Sweet replied, in effect, 'Instead of three circles in a plane, imagine three balls lying on a surface plate. Instead of drawing tangents, imagine a cone wrapped around each pair of balls. The apexes of the three cones will then lie on the surface plane. On top of the balls lay another surface plate. It will rest on the three balls and will be necessarily tangent to each of the three cones, and will contain the apexes of the three cones. Thus the apexes of the three cones will lie in both of the surface plates, hence they must lie in the intersection of the two plates, which is of course a straight line.'

[Graham, 1963, problem 62]

482. This is the solution given by Harry Lindgren and Greg Frederickson, in their marvellous book *Recreational Problems in Geometric Dissections and How to Solve Them*.

Since the final assembled star must have an edge length $\sqrt{3}$ times that of the original stars, it is natural to join the vertices of a smaller star to the centre, because that radius is indeed $\sqrt{3}$ times the edge length of the smaller stars.

[Lindgren and Frederickson, 1972, pp. 104–5]

483. The hidden feature is that the side of the square is equal in length to the line joining a vertex of the dodecagon to the next-vertex-but-three. Such simple relationships between polygons of equal

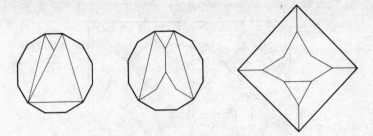

area invariably indicate that a mutual dissection is relatively simple.

These are two solutions, each in six pieces. The first has the remarkable property that it can be executed with no measurement at all, simply by joining suitable vertices of the dodecagon.

[Lindgren and Frederickson, 1972, p. 41–2]

484. The smallest in area, and measured by the length of the shortest side, is 3 × 7. This is one solution with attractive symmetry: the knight starts at 1 and moves to 2, 3, . . . in sequence.

9	6	3	20	17	12	15
4	1	8	11	14	21	18
7	10	5	2	19	16	13

For the tour to have rotational symmetry, the board must be at least 6 × 6. There are five such tours on the 6 × 6 board, of which this is one:

[Kraitchik 1955, p. 263]

485. The minimum solution requires seventeen moves.

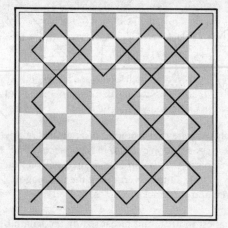

[Domoryad, 1963, p. 118]

486. Right-handed.

487. There are eighty magic hexagons, in twelve of which the outer vertices also sum to the magic constant, 26.

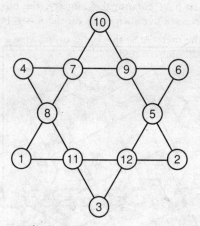

[Gardner, 1978b, p. 58]

488. 'Eight swings are enough to reverse the two bookcases. One solution: (1) Swing end B clockwise 90 degrees; (2) swing A clockwise

30 degrees; (3) swing B counterclockwise 60 degrees; (4) swing A clockwise 30 degrees; (5) swing B clockwise 90 degrees; (6) swing C clockwise 60 degrees; (7) swing D counterclockwise 300 degrees; (8) swing C clockwise 60 degrees.'

[Gardner, 1977, p. 180]

489. Slide the horizontal match to one side, by half its length, and then move the unattached match to form the remaining side of the cocktail glass, which now contains the cherry.

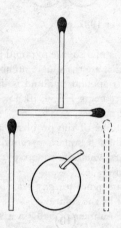

490. Assemble the matches in this order, using long kitchen matches in preference to the shorter kind.

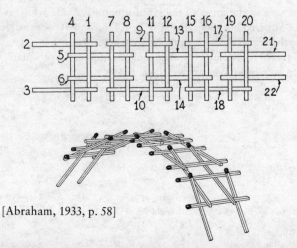

[Abraham, 1933, p. 58]

491. Each symbol is one of the digits 1 to 9, with its reflection in a vertical mirror. The numbers are arranged to form a magic 3 × 3 square, so the symbol in the empty cell is an 8 with its reflection.

[*Plus*, No. 6, Summer 1987, p. 7]

492. The number of balls in a square pyramid is the sum of the layers from top to bottom, which are the square numbers, 1, 4, 9, 16, 25 . . .

The number on the triangular pyramid is likewise the sum of the triangular numbers, there being 1, 3, 6, 10, 15, 21, 28 . . . balls in each layer from the top downwards.

To satisfy the conditions of the problem, a triangular pyramidal number is one more than a square pyramidal number. By simple addition, this first occurs when the square pyramid contains 55 balls and the triangular pyramid, 56.

Since it does not occur again for any reasonably small sizes of the two pyramids, this must be the solution.

(The formula for the number of balls in a square pyramid of side n is $\frac{1}{6}n(n + 1)(2n + 1)$, and for the triangular pyramid, $\frac{1}{6}n(n + 1)(n + 2)$.)

493. The sum of the areas of the given squares is 31152 which is not a perfect square, but does equal 177 × 176, and it is possible to assemble them into the near-square shown opposite.

It is possible to assemble distinct squares into one large square, but at least twenty-one pieces are required.

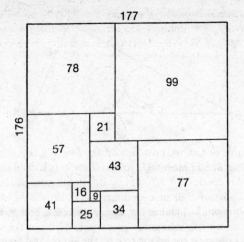

494. Yes, you can, because one of the books has to contain no words at all, and another contains just one word. If the number of books in the library is N, which is greater than the number of words in any of them, the numbers of words must be $N - 1$, $N - 2$, $N - 3$ ... all the way down to 3, 2, 1, 0.

Since the problem states that there are 'books' in the library, you can safely assume at least two books, but not more.

[Mensa, 1975, problem 27]

495. Extremely high. The number of molecules of air breathed by Archimedes during his lifetime is vastly greater than the number of litres of air in the earth's atmosphere. Assuming that the air breathed by Archimedes has had plenty of time over more than 2000 years to become thoroughly mixed with the whole atmosphere, and even allowing for a proportion of the air which has been taken into the ocean or used up in chemical reactions and remained locked away, it is still highly likely that you are breathing some of Archimedes' air.

496. In half. This can be seen by a typical bit of argument by analogy.

Imagine 2 kg at A and 1 kg at each of B and C. For the purposes of finding the centre of gravity of all three weights, the latter two can be replaced by a 2 kg weight at A', and the centre of gravity will lie half way between the two 2 kg weights, in other words, at Y.

Now think of the weights at A and C replaced by a 3 kg weight at

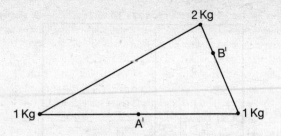

B'. The centre of gravity now lies on the line B'B, dividing it in the ratio, however, of 1:3. So their intersection Y is indeed the centre of gravity, and it divides AA' in half and BB' in the ratio 3:1.

497. Only one set, 1, 2 and 3: $1 + 2 + 3 = 1 \times 2 \times 3 = 6$.

498. He ties the rope round the tree on the shore, and then carries the rope on a walk round the island. As he passes the halfway mark, the rope starts to wrap around the tree on the island, and when he reaches his starting point he ties the other end of the rope to the tree on the shore and pulls himself across on the rope.

[Gardner, 1983, Chapter 8, problem 35]

499. This problem can be solved, appropriately, by use of the pigeon-hole principle.

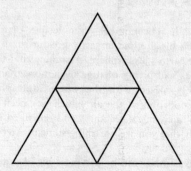

Dissect the field into four equilateral triangles, each with sides of 50 metres. The pigeon-hole principle says that if five objects are placed in only four boxes, then one of the boxes contains at least two objects.

In the present case, one of the triangular 'quarters' contains at least two droppings, which will be at most 50 metres apart (allowing for the possibility that one or more droppings might lie exactly on one of the dividing lines).

500. Consider the coordinates of a pair of lattice points, and in particular whether they are odd or even. The mid-point of the line joining them will only be itself a lattice point if they match in parity. But there are only four possible patterns of odd-and-even. A particular point may have coordinates which are odd-odd, even-even, even-odd, or odd-even.

Therefore if five points are taken, then by the pigeon-hole principle, at least two of them must have the same pattern of coordinates, and their mid-point will lie on the lattice.

[Larsen, 1983]

501.

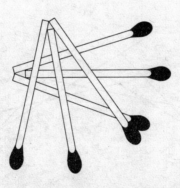

502. If More starts, then Less has the perfect strategy of taking enough matches to bring his take and More's last take up to the magic number four. So if More takes two, Less takes two, but if More takes three, then Less takes one, and so on.

This ensures that after five turns each, twenty matches will have gone, and More loses by taking the last.

When it is Less's turn to start, he can only hope that More does not know the winning strategy and will allow Less sooner or later to take a number which brings the total taken up to a multiple of four. If More does know the trick, then at least they will win alternate games.

503.

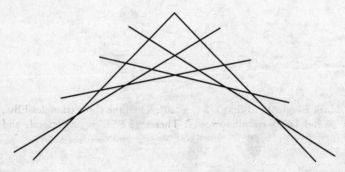

504. The maximum possible number with six lines is 20 which is just the number of ways of choosing three lines out of six to be the sides of the triangle.

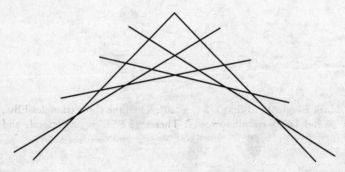

505. These are probably the simplest solutions, and are easily sketched on hexagonal or triangular paper:

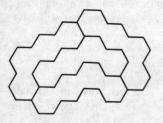 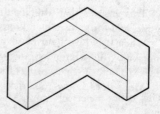

[Gardner, 1989]

506. Angle BDE = 30°. Angle-chasing is not sufficient to find this angle. A complicated general formula can be found by repeated use of the sine rule, or, in this case, a simple geometrical construction can be exploited.

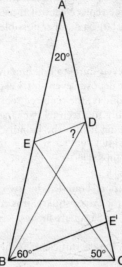

Mark E′ on AC so that E′BC = 20°. Then the three triangles EBC, BE′C and DE′B are all isosceles. Therefore BEE′ is equilateral, and triangle EE′D is isosceles. But DE′E = 40°, and so BDE + 40° = 70° and BDE = 30°.

[Tripp, 1975]

507. Suppose you start with coin 1. Count 1, 2, 3 and turn 4 tails up. Count 6, 7, 8 and turn 1 tails up. Count 3, 4, 5 and turn 6 tails up. Count 8, 1, 2 and turn 3 tails up. Count 5, 6, 7 and turn 8 tails up. Count 2, 3, 4 and turn 5 tails up. Count 7, 8, 1 and turn 2 tails up.

508. 'Let N be the smallest integer. The product is then

$$N(N + 1)(N + 2)(N + 3) = (N^2 + 3N)(N^2 + 3N + 2)$$
$$= (N^2 + 3N + 1)^2 - 1$$

This is not a perfect square since two positive squares cannot differ by 1.'

[Dunn, 1980, p. 92]

509. 'Clearly 1 would not appear as a factor, and any 4 could be replaced by two 2's, without decreasing the product. And if one of the factors were greater than 4, replacing it by 2 and $n - 2$ would yield a larger product. Thus the factors are all 2's and 3's. Moreover, not more than two 2's are used, since the replacement of three 2's by two 3's would increase the product. The largest number possible is therefore $3^{32} \times 2^2$.'

[Dunn, 1980, p. 84]

510. 'He must pick up seven shirts to tide him over until the following Monday. Hence he must deposit seven shirts each Monday. Counting the shirt he wears on Monday, the required total is fifteen. (Note that he cannot get by with only fourteen by exchanging his Monday shirt for a clean one and turning it into the laundry, as he will be caught short the following Monday.)'

[Dunn, 1983, problem 84]

511. Because the overlap is bounded by two lines at right-angles meeting at the centre of the larger square, it is equal to one quarter of the larger square in area, or 25 square inches.

512. A strip just 8 inches long is sufficient, as the diagram shows:

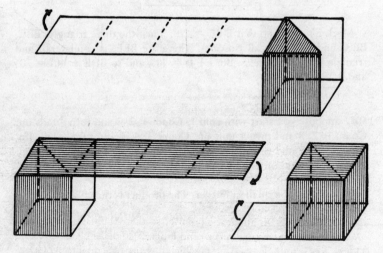

The strip is folded at 45° to bring it to the left (top figure) and then back across the triangle, and so continues in a clockwise direction.

[Madachy, 1966, p. 124, problem 23]

513. The order makes no difference, the final price is the original price multipled by (95/100) × (90/100) × (80/100), or 68.4 per cent.

514. The thickness will be $2^{50} \times \frac{1}{10}$ mm, which is rather more than 70,368,681 miles, 'or more than two-thirds of the distance from the earth to the sun'.

[Tocquet, 1957, p. 109]

515. The simplest answer is 9,876,543,210 − 0,123,456,789 = 9,753,086,421.

516. $99066^2 = 9,814,072,356.$

517. Label the rectangle ABCD. Fold AC on to AB, to form the edge CE, and then fold BD on to CE, to form the crease FG. Finally fold BD towards A, so that the new crease passes through F and so that BDA is a straight line.

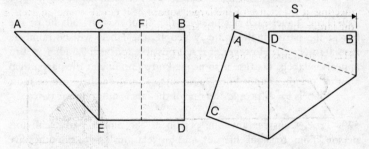

AB is then equal in side to the square whose area is equal to the original rectangle. This is Abraham's proof:

'Let AB = x = y and AC = $x - y$. Then area of ABCD = $x^2 - y^2$ and AB − AC = $2y$.

'*First* – Fold the short edge AC up on to the line AB as shown in the first figure. CB is therefore equal to $2y$.

'*Second* – Fold BD over to CE on line F and open out these folds. BF is therefore equal to Y, and AF = x.

'*Third* – Fold DB about the point F, so that ADB is a straight line. We now have the triangle AFB, and DBF is a right-angle. As AF = x, BF = y, S is the side of a square which has an area equal to ABCD.'

[Abraham, 1933, problem 99]

518. The pieces will only form a hollow triangle within a square:

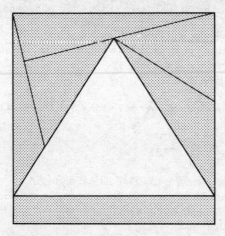

519. True. Label each person with the number of friends he or she has at the party. If there are N people, the labels will be numbers from 0 to $N - 1$. However, it is not possible for both 0 and $N - 1$ to appear as labels, because if someone knows no one at the party, then another person cannot know everyone. Therefore there are at most $N - 1$ labels for N people and one of the labels must appear twice.

520. True. Consider any one person at the dinner party (call this person 'Tom' for convenience), and his relationships to the other five present. Of these five, either at least three are friends of Tom, or at least three are strangers to him.

If they include three friends of Tom, then either these three are all strangers, or one pair are mutual friends and form with Tom a group of three mutual friends. Similarly, if they include three strangers, they are either mutual friends, or two of them and Tom are mutual strangers.

521. In this diagram the intersections of the paths have been marked with the numbers of ways in which Lady Merchant can reach the intersection. For each intersection this number is the sum of the numbers at the previous intersections from which the intersection can be reached by walking along the side of one plot.

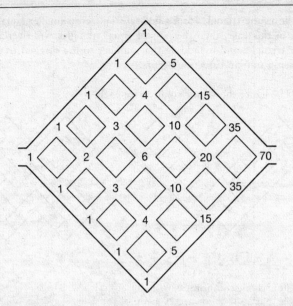

The numbers, in fact, are the numbers in Pascal's Triangle, and the summer house can be reached in a total of seventy different ways.

522. Any triangular pattern of dots can be divided into a 'square' (which does contain a square number of dots, although it is skew) and two triangles.

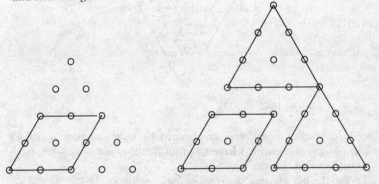

If the 'square' includes the middle dot on one side, as in the middle figure, then the next smaller square will leave two larger triangles overlapping, as in the right-hand figure, and the addition of one extra dot allows the triangles to be separated. (If even smaller squares are taken in this case, then the pair of equal triangles will overlap in a larger triangle, and by adding that number to the original triangular number, a similar dissection occurs.)

523. The figure shows one solution for each.

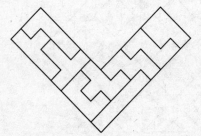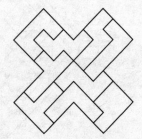

[Golomb, 1965, p. 26]

524. This is one solution:

[Golomb, 1965, p. 132]

525. These solutions were contributed by Bob Newman to *Games and Puzzles* magazine. I have no proof that they are minimal, though I suspect that they are.

On the left are individually locking 21-ominoes. In the middle are 14-ominoes which interlock as a tessellation, but not individually, and on the right are 12-ominoes with the same property, but one half of the tiles have to be turned over.

[Wells, 1979, problem 61]

526. $1,000,000,000 = 10^9 = 2^9 \times 5^9 = 512 \times 1953125$.

Any power of 10 can be expressed as the product of the same. powers of 2 and 5, but the latter usually themselves contain at least one zero. Two larger products which are zero-free are 10^{18} and 10^{33}.

[Ogilvy, 1966, p. 89]

527. For two children in general there are four equally likely events: boy–boy, girl–girl, boy–girl and girl–boy. Since boy–boy is ruled out, the chance of girl–girl is $\frac{1}{3}$.

Taking the same four events in older–younger order, both girl–boy and girl–girl are ruled out, so the probability of two boys is $\frac{1}{2}$.

[Kordemsky, 1972, problem 236]

528. Mrs Tabako should place one cultured pearl in one jar, so that if Mr Tabako chooses that jar his chance of success is 100 per cent, and place all the remaining pearls, forty-nine cultured and fifty natural, in the other jar, so his chance will be 49/99.

[Morris, 1970, p. 136]

529. Curiously, the answer depends on what means you choose to define the random chord: the qualification 'random' by itself is not sufficient to force a unique solution.

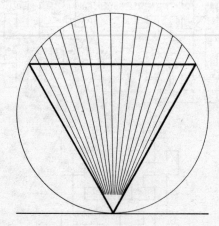

Consider first that the positions at which the chord meets the circle are not important but the angle at which it meets the circle is, so consider the chords through a given point on the circle. Comparing their lengths with the equilateral triangle with a vertex at the same point, the chord will be longer than the triangle-side if it falls within the central angle of 60°, and the chance of this is 60/180 = $\frac{1}{3}$.

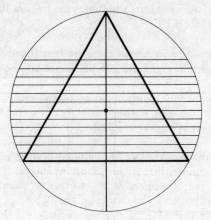

Next, consider the idea that the direction of the chord does not matter – it is sufficient to consider all possible positions of the chord parallel to a given direction. If the chord falls within the central band it will be longer than the triangle side, otherwise not. But then length XY is one half of the diameter of the circle, so in this case the probability is $\frac{1}{2}$.

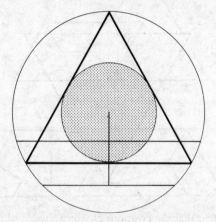

A random chord can have its centre anywhere at all within the circle, but only a chord whose centre lies within the circle inscribed in the equilateral triangle will be longer than the triangle side. So the required probability is the ratio of the areas of the smaller circle to the larger, which is $1:2^2$, or $\frac{1}{4}$.

[Northrop, 1960, pp. 169–70. Other methods of interpreting the qualification 'random' produce more different answers. Thus Hunter and Madachy, 1963, p. 102, produce the answer $\frac{3}{2}$]

530. The figure can be divided into any number of identical pieces (including two) by repeating the matching shapes of the ends.

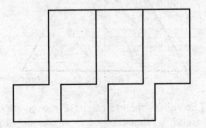

[After Eastaway, 1982, p. 103]

531. This shape is the only known pentagonal reptile of order 4, that is, which divides into identical quarters.

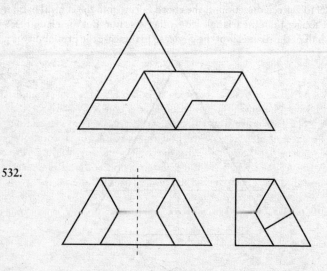

532.

Divide the first figure about the vertical line of symmetry, and bisect its other two tiles similarly, to get the second figure.

533. The process described is called Kaprekar's process, after D. R. Kaprekar, an enthusiastic Indian amateur mathematician who has been publishing his results in number theory for many years.

The result of the process, when sufficiently repeated, is always the number 6174, called Kaprekar's constant. This is just one example, starting with the number 4527:

$$7542 - 2457 = 5085$$
$$8550 - 0558 = 7992$$
$$9972 - 2799 = 7173$$
$$7731 - 1377 = 6354$$
$$6543 - 3456 = 3087$$
$$8730 - 0378 = 8352$$
$$8532 - 2358 = 6174$$
$$7641 - 1467 = 6174 \text{ and the calculation repeats.}$$

534. 2 metres. The rollers themselves have moved forward 1 metre, relative to the ground, so that the points at the top of the rollers on which the slab originally rested are 1 metre ahead. The slab has moved an equal distance relative to the rollers, making a total movement of 2 metres.

[Northrop, 1960, p. 53]

535. He started at house No. 5 and went 5–10–1–9–2–8–3–7–4–6, a total of 4900 metres.

[After Gardner, 1971, p. 235]

536. Neither. The distance between the heads does not change. Think about it by imagining that only one bolt moves about the other. It will either move towards the other or move away depending on the direction of the screw on the bolt and the direction of rotation.

No matter what happens with the bolt you are holding, the opposite will happen when you hold the first bolt still and move the second around it. Therefore, when you 'twiddle' them, the two motions will cancel each other out.

[Lukacs and Tarjan, 1982, p. 170]

537. No. Here is a counter example.

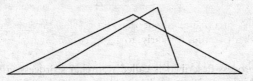

538. She takes eight stockings and is assured of at least one pair.

539. Extraordinary to relate, it is impossible to play a man as far as the fifth rank, though any army of twenty men can send a man to the fourth rank, to X, if the men are arranged as shown.

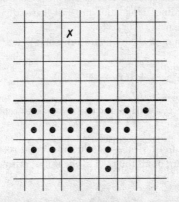

Trial and error may well convince the reader of this impossibility, which can be proved by the following beautiful argument, which depends on picking a target square on the fifth rank, to which you hope to despatch a man, and labelling it and surrounding squares as in the figure below.

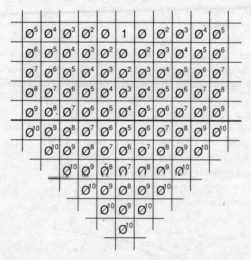

Here, the letter $\emptyset$ stands for $\frac{1}{2}(\sqrt{5} - 1)$, so that $\emptyset$ has the property that $\emptyset^2 + \emptyset = 1$, and – this is the point of the labelling – every legal move consisting of a solitaire jump *leaves the sum total of the values of all the squares occupied unchanged*. For example, if a man occupying $\emptyset^7$ jumps over a man occupying $\emptyset^6$ it will be into a square labelled $\emptyset^5$, and $\emptyset^7 + \emptyset^6 = \emptyset^5$.

Now consider the total value of all the men in the initial army below the starting line. This value must at least equal 1 if one man is finally to reach square 1.

However, the sum of all the men in the first row below the starting line is less than

$$(\emptyset^5 + \emptyset^6 + \emptyset^7 + \ldots) + (\emptyset^6 + \emptyset^7 + \emptyset^8 + \ldots)$$

$$= \frac{\emptyset^5}{1 - \emptyset} + \frac{\emptyset^6}{1 - \emptyset} = \frac{\emptyset^5}{\emptyset^2} + \frac{\emptyset^6}{\emptyset^2} = \emptyset^3 + \emptyset^4 = \emptyset^2$$

Similarly, the sum of all the men in the second row is less than $\emptyset^3$. Continuing, the sum of all the men in the initial army is less than

$$\emptyset^2 + \emptyset^3 + \emptyset^4 + \ldots = \frac{\emptyset^2}{1 - \emptyset} = 1$$

Therefore no man can reach the fifth rank. If, however, just one cell, no matter where, is allowed to be occupied by two men, then the fifth rank is within reach.

[Beasley, 1989, pp. 86–7. John Beasley, who has also written *The Ins and Outs of Peg Solitaire*, comments that 'Solitaire offers many lovely problems, but this is one of the loveliest']

540. Yes, it is. Here is a square surrounding seventeen points.

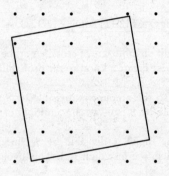

The general answer is 'Yes' also. An intuitive proof can start with the idea that if you have a square which surrounds exactly $N - 1$ points, then by keeping the same centre and orientation you can increase the size of the square until either one point lies on its perimeter, in which case a sufficiently small further increase will bring that point inside without putting any more points on the perimeter, or several points appear on the perimeter at once, in which case you retreat slightly, give the square a very slight movement, and increase the size again until this time just one new point appears on the perimeter.

[Honsberger, 1973, p. 121, where Honsberger gives Browkin's watertight proof of this conclusion]

541. James's book was published in 1907, two years after Einstein published his first paper on his special theory of relativity, so it is appropriate that the answer is relative – it depends what you mean by the verb 'circle'.

It is a well-known fact that the moon always turns its same face to us, and that its distance from the earth was once less than it is now.

We can imagine, therefore, a planet with a moon which always faces it, and which circles it in phase with the planet's own rotation. Such a moon would always face the planet at the same point in the sky.

Relative to the position of the planet in space, the moon is circling the planet, and the planet is also circling the moon, relative to the moon's path through space, but the planet and moon are never behind each other because of their own matching rotations. Therefore, if your idea of circling includes the idea of 'going behind' then they do *not* circle each other.

The situation of the hunter and the squirrel is basically the same, except that the tree does not move through space. Relatively simple?

542. 12128 farthings = £12 12s 8d, or £12/12/8.

543. There are eight possible sets of crossings at the three intersections and only two of these create a knot, so the probability is 1/4.

544. 72 = 9 × 8 so the amount is divisible by 8 and 9. Thousands are always divisible by 8, so 79– is a multiple of 8, and so the last digit is 2.

The sum of the digits is divisible by 9, since the number is a multiple of 9, and so the first digit is 3. The invoice was for £367.92 and each bird cost £5.11.

[Beiler, 1964, p. 302, problem 73]

545. 'On the first day, the chauffeur was spared a 20-minute drive. Thus Mr Smith must have been picked up at a point which is a 10-minute drive (one way) from the station. Had the chauffeur proceeded as usual, he would have arrived at the station at exactly 5 o'clock. The 10-minute saving means that he must have picked up Smith at 4.50. Thus Smith took 50 minutes to walk what the chauffeur would take 10 minutes to drive. From this we see that the chauffeur goes five times as quickly as Smith.

'Now, on the second day, suppose that Smith walks for $5t$ minutes. The distance he covers, then, would take the chauffeur only t minutes to drive. Accordingly, Smith was picked up this time at t minutes before 5 o'clock, that is, at $60 - t$ minutes after 4 o'clock. However, starting at 4.30 and walking for $5t$ minutes, Smith must have been picked up at $30 + 5t$ minutes after 4 o'clock. Hence $30 + 5t = 60 - t$, and $t = 5$. Therefore the chauffeur was spared a 5-minute drive (each way), providing a saving of 10 minutes this time.'

[Honsberger, 1978, problem 6]

546. The train takes 30 seconds to travel 1 km, plus 3 seconds for the complete train to pass any point, making a total of 33 seconds.

547. The paddle, being unpowered, does not move relative to the water, and since the boat has a constant speed relative to the water, regardless of that speed, it will take the boat as long to get back to the paddle – 10 minutes – as it did to get away from it. Since the paddle (and the water) moved 1 mile in that total period of 20 minutes, the current must have been 3 mph.

[Graham, 1968, problem 24]

548. It is not true that if one sees two sides the other will see three. In most positions, they will each see two sides. A quick sketch will show that the chance of seeing three sides at once is rather small, and only approaches fifty-fifty as the spy recedes to an infinite distance from the building.

549. Let the longer candle be x inches originally, and burn at r inches per hour. Then the shorter candle was originally $x - 1$ inches long. Let s denote its rate of burning. At 8.30, the longer candle has burned for 4 hours and the shorter for $2\frac{1}{2}$ hours, and they are the same length:

$$x - 4r = (x - 1) - \frac{5s}{2}$$

Also, the longer candle is consumed in 6 hours and the shorter in 4 hours, so:

$$6r = x \quad \text{and} \quad 4s = x - 1$$

From these equations, $r = 1\frac{1}{2}$, $s = 2$ and $x = 9$, so the candles were 9 inches and 8 inches long.

[NCTM, 1965, problem 76]

550. The second box from the right was originally in the middle.

'If the boxes are thought of as dice with numbers on them (conventionally arranged so that pairs of numbers on opposite faces add up to 7) the sum of the numbers for the upper, front and right-hand faces of any of the dice is an odd or even number according to its orientation; for instance the diagram below (assuming the face with the A has one pip on it) we have

$$1 + 2 + 3 = 6 \quad \text{(even)}$$
$$2 + 6 + 3 = 11 \quad \text{(odd)}$$
$$4 + 2 + 1 = 7 \quad \text{(odd)}$$

In this way we can call the orientation of a box either *odd* or *even*. The class of orientation changes each time a box is tipped over on an edge. This diagram shows this for two cases – where the first box is tipped over backwards, or to the right. The result follows in the general case from the fact that each tipping over of the box leaves two of the previous numbers to be considered again, while the third is replaced by its difference from 7 and thus becomes odd if it was even before, and conversely. So the sum of the upper, front and right-hand faces changes in similar fashion – the box goes from an odd orientation to an even one, or vice versa. It is not in fact difficult to show that successive tipping operations can indeed produce any even orientation on any square of unchanged colour and any odd orientation on any square of altered colour.

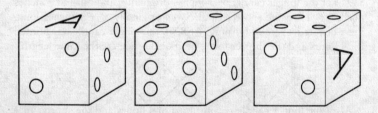

'If all the boxes in position A are taken to have even orientations, only the second from the left in position B has an odd one; so the latter must have been tipped over an odd number of times, and the rest an even number of times.

'Now we assume that all this takes place on a floor which is chequered like a chessboard, where each box exactly covers one square. Every time a box is tipped over, this changes the colour of the square which it covers; so each box stands on one colour in all its even orientations, and on the other colour in all its odd ones. Suppose that the first, third and fifth boxes are on black squares in position B; then they were on black squares in position A also. The second box in position B, which is now on a white square, must in view of its new odd orientation have been on a black square, earlier, in position A. So we recognize these four boxes as those which occupied four like-coloured squares in position A – i.e. the outer ones. The second box from the right in position B, with an even orientation on a white square, stood also on a white square in position A, and so was in the middle.'

[Sprague, 1963, problem 3]

551. Call the trains A and B. A runs one third of its length in 1 second and B runs one quarter of its length, so they separate by $\frac{1}{3} + \frac{1}{4} = \frac{7}{12}$ of the length of either in 1 second. To pass each other, they must separate by twice the length of either, which will take $2 \times \frac{12}{7} = \frac{24}{7} = 3\frac{3}{7}$ seconds.

[Workman, 1920, p. 406, exercise 7]

552. $4^3 = 64$. In general, if two sides of the original triangle are divided into N parts, there are N^3 triangles in the figure, made up of N^2 triangles whose sides include the base of the original triangle, and $2 \times \frac{1}{2}N(N-1) \times N$ whose sides do not.

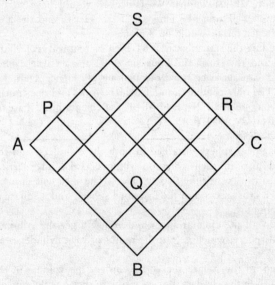

The answer can also be found by visualization: break the base of the triangle, and slightly separate the lines going into each corner. Continue to bend the two halves of the base and to separate the same lines until the figure is transformed into a grid of squares. Each original triangle to be counted corresponds to a rectangle, at least one of whose sides lies along the edges AB and BC. The number of such rectangles is the total number of rectangles in the figure, less the rectangles in the top square PQRS. The first number of rectangles is T^2_N where T_N is the Nth triangle number, $\frac{1}{2}N(N-1)$, and the second number is T^2_{N-1}. The differences between the squares of the triangular numbers, are the cubes of the integers: $T^2_N - T^2_{N-1} = N^3$.

[Wells, 1983–6, Series 1, problem 28]

553. He asks either of the warders, 'Does the warder who is guarding the road that leads to freedom tell the truth?' If the warder he asks replies 'Yes', he goes through that warder's door. If the warder replies 'No', he walks to freedom through the other door.

554. Consider the puzzle with statements 1–5 only, and denote '2 is true' by 2T, and so on.

2T implies 5T implies 4F (since 4T, 5T would make 4F).

4F implies two consecutive true statements, which is impossible.

Therefore 2F. 2F implies 4T (since 4F would have to be true), which implies 3F and 5F. Thus 1F 2F 3F 4T 5F, and the answer to this shortened puzzle would be '4 alone'.

But adding the statement 6, if 6T then its removal *would* affect the answer, and therefore 6F. Thus since 6F, the answer *cannot* be 4 alone. 2T can again be ruled out making 2F. Hence again 4T and 3F and 5F. The only combination left which works and does not give the answer '4 alone' is 1T 2F 3F 4T 5F 6F. Statements 1 and 4 are true.

[Eastaway, 1982, p. 15]

555. Imagine a sphere inscribed in both the cylinders, and therefore inscribed also in their common volume. Take any vertical slice parallel to the axes of the two cylinders. The slice will show a circle, which is a cross-section of the inscribed sphere, and a square circumscribed about it.

Add up all the circular slices and you will get the volume of the inscribed sphere, which is $\frac{4}{3}\pi$ if the radius of each cylinder is taken to be 1 unit.

Add up all the square slices and you get the volume of the solid common to both cylinders. But each square slice has the same ratio to its inscribed circle, 4 to π, and so the volume of the common solid has the same ratio to the volume of the inscribed sphere.

The common solid therefore has volume $\frac{16}{3}$. It is notable that it is a rational number, and does not involve π.

[Gardner, 1977, p. 176]

556. The four points of intersection of circle E, which needs no proof, and the points A, B, C and D.

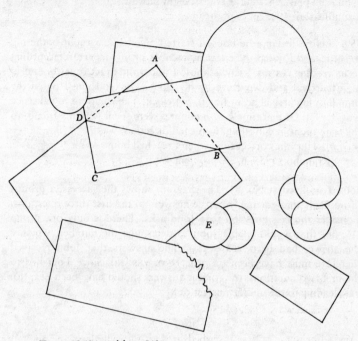

[Barr, 1969, problem 51]

557. Three boys and three girls each received one $\frac{1}{2}$ cent and two $\frac{1}{3}$ cent buns. There could also have been only one boy and one girl, but then the purchase could have been made in ten ways instead of the single way required.

[Beiler, 1964, p. 298, problem 32]

558. Black can only have moved his knights, which are back on their original squares (having possibly swopped places), and his rooks which can only have moved to the original position of a knight and back again, and so Black has made an even number of moves. Similarly White's knights and rooks have only made an even number of moves. Therefore, to account for the odd single move by the White pawn, the White King or Queen must have made an odd number of moves between them, and this is most briefly accomplished by the White King moving seven times, for example Ke1–f2–e3–f4–g4–g3–

f2–e1, in which the move to g4 ensures the odd number of moves before return to this starting square.

So each player has made at least eight moves.

[Beasley, 1989, p. 77]

559. 'Before leaving the house I started my clock, without bothering to set it, and I noted the exact moment A of my departure according to its reading. At my friend's house I noted the exact times, h and k, of my arrival and departure by his clock. On returning I noted the time B of my arrival according to my clock. The length of my absence was $B - A$. Of that time $k - h$ minutes were spent with my friend, so the time spent in travelling, t, in each direction, was $2t = (B - A) - (k - h)$. Thus the correct time when I reached home was $k + t$.'

[Kraitchik, 1955, p. 39, problem 49]

560. Tom lived at No. 81. The diagram shows the successive groups of numbers that emerge from the answers to the first three questions. Consider the last question that John asks. There is only one group (16, 36) that would enable John to determine the number, whether the answer had been YES or NO. The answers that John received therefore must have been (in order) NO, YES, NO. Since Tom lied to the first two questions, the correct answers should have been YES, NO, YES, leading to the unique answer of 81.

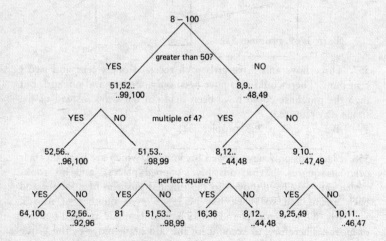

[*Plus*, No. 7, Autumn 1987, p. 13, problem 4]

561. Yes. 'sister' must mean 'step-sister'. Mr Jones's mother and Mr Smith's father, having divorced their original spouses, remarry each other and have a daughter, Flo, who is step-sister to each of the men who are completely unrelated.

    [After Kasner and Newman, 1949, p. 183]

562. Suppose B's statement is true. Then A's statement would be true and C would be a Pukka. But this is not possible, for all three would then make true statements.

    Therefore B's statement is false, and A is not a Pukka.

    Therefore neither A nor B is a Pukka, and so C is a Pukka.

    Therefore A's statement is true, and so A must be a Shilli-Shalla, and B is the Wotta-Woppa.

    Conclusion: A is a Shilli-Shalla; B is a Wotta-Woppa; C is a Pukka.

    [Emmett, 1976, problem 21]

563. The squares, as assembled here, wrap round so that the left and right edges match.

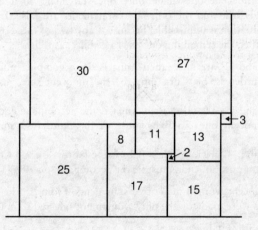

564. 'Let a be the amount A had and b be the amount B had before A and B bet. Then, from [1], after they bet A had $2a$ and B had $b - a$.

    'Let c be the amount C had before he bet with B. Then, from [2], after B and C bet, B had $(b - a) + (b - a)$ or $2b - 2a$, and C had $c - (b - a)$ or $c - b + a$.

    'Then, from [3], after C and A bet, C had $(c - b + a) + (c - b + a)$ or $2c - 2b + 2a$, and A had $2a - (c - b + a)$ or $a - c + b$.

'From [4], $a - c + b = 2b - 2a$ and $a - c + b = 2c - 2b + 2a$. The first equation yields: $b = 3a - c$, and the second equation yields: $3b = a + 3c$. Multiplying the first of these latter equations by 3 and adding the two equations yields: $6b = 10a$ or $b = \frac{5}{3}a$. Substitution for b yields: $c = \frac{4}{3}a$.

'So A started with a cents, B with $\frac{5}{3}a$ cents, and C with $\frac{4}{3}a$ cents.

'From [5], a cannot be 50 cents because then B and C would have started with fractions of a cent, and $\frac{4}{3}a$ cannot be 50 cents because then A and B would have started with fractions of a cent. So $\frac{5}{3}a$ is 50 cents and B *is the speaker.*

'In summary, A started with 30 cents, B started with 50 cents, and C started with 40 cents.'

[Summers, 1968, problem 8]

565. 'Because each statement refers to a different woman, the innocent one did not make all three statements; otherwise, she would have spoken of herself, contradicting [I]. So the innocent one made either one statement or two statements, from [II].

'If the innocent one made only one statement, then only that statement is true and the other two statements are false, from [III]. But this situation is impossible, because if any two of these statements are false, then the remaining one has to be false, as follows:

(a) If [1] and [2] are false, then Anna is the accomplice and Babs is the murderess. So Cora must be the innocent one, making [3] false.
(b) If [1] and [3] are false, then Anna is the accomplice and Cora is the innocent one. So Babs must be the murderess, making [2] false.
(c) If [2] and [3] are false, then Babs is the murderess and Cora is the innocent one. So Anna must be the accomplice, making [1] false.

So the innocent woman made two statements. From [I], the two true statements were made by the only woman not referred to in these two statements:

(d) If statements [2] and [3] are true, they were made by Anna. Then Anna is the innocent one. But [1], being false, identifies Anna as the accomplice. This situation is impossible.
(e) If statements [1] and [3] are true, they were made by Babs. Then Babs is the innocent one. But [2], being false, identifies Babs as the murderess. This situation is impossible.
(f) So statements [1] and [2] are true, and were therefore made by

Cora. Then Cora is the innocent one. The falsity of [3] is consistent with this conclusion. Since Cora is the innocent one and [1] is true, Babs is the accomplice. Then *Anna is the murderess*. [2], being true, is consistent with this conclusion.'

[Summers, 1968, problem 27]

566. '$F \times ABCDE = GGGGGG$

$F \times ABCDE = G \times 111111$

Of the numbers 2 through 9, 111111 is divisible exactly by only 3 and 7.

$F \times ABCDE = G \times 3 \times 7 \times 5291$.

'If G is a multiple of F, then $ABCDE$ would be a number containing the same digit 6 times. So G is not a multiple of F.

'Then: (a) F does not equal zero, otherwise G would equal zero and, therefore, would be a multiple of F.

(b) F does not equal 1, otherwise G would be a multiple of F.

(c) F does not equal 2, otherwise G would have to be a multiple of 2 (for an exact division) and, therefore, a multiple of F.

(d) F does not equal 4, otherwise G would have to be a multiple of 4 (for an exact division) and, therefore, a multiple of F.

(e) F does not equal 8, otherwise G would have to be 8 also (for an exact division) and, therefore, a multiple of F.

(f) F does not equal 5, otherwise G would have to be 5 also (for an exact division) and, therefore, a multiple of F.

(g) If $F = 3$, then: $ABCDE = G \times 7 \times 5291 = G \times 37037$. The presence of a zero in 37037 indicates that the product of any single digit times this number will result in duplicate digits for $ABCDE$. So F does not equal 3.

(h) If $F = 6$, then: $ABCDE \times 2 = G \times 7 \times 5291 = G \times 37037$. G, then, must be a multiple (M) of 2, that is $G/2 = M$. Then: $ABCDE = M \times 37037$. By the reasoning in (g), F does not equal 6.

(i) If $F = 9$, then: $ABCDE \times 3 = G \times 7 \times 5291 = G \times 37037$. So G must be a multiple (M) of 3, that is $G/3 = M$. Then: $ABCDE = M \times 37037$. By the reasoning in (g), F does not equal 9.

(j) So $F = 7$. Then: $ABCDE = G \times 3 \times 5291 = G \times 15873$. Since there are seven different digits involved, G does not equal 1, 5, or 7. Since $ABCDE$ contains only

five digits, G does not equal 8 or 9. Since F does not
equal 0, G does not equal zero. So G equals 2, 3, 4, or 6.
The four possibilities are:

$$F = 7, G = 2, ABCDE = 31746$$
$$F = 7, G = 3, ABCDE = 47619$$
$$F = 7, G = 4, ABCDE = 63492$$
$$F = 7, G = 6, ABCDE = 95238$$

Only the last one of these possibilities results in seven different digits.
The multiplication, then, is

$$\begin{array}{r} 9\ 5\ 2\ 3\ 8 \\ \times\qquad 7 \\ \hline 6\ 6\ 6\ 6\ 6\ 6 \end{array}$$

and G represents 6.'
    [Summers, 1968, problem 34]

567. 'Sales: third week, x cars; second week, y; first week, $(56 - x - y)$.
From data $x^2 - (55 - y)x - 2y^2 + 57y - 56 = 0$.
This is an "indeterminate equation of the second degree", i.e. two
unknowns, including the square of one or both. Solution depends on
the fact that both unknowns are whole numbers. We treat it as an
ordinary quadratic, in this case in x:

$$x = \frac{(55 - y) \pm \sqrt{(9y^2 - 338y + 3249)}}{2}$$

As x and y are whole numbers, the expression under the square-root
sign must be square and positive.

So let $9y^2 - 338y + 3249 = k^2$

where k is any whole number that will satisfy the square condition.
Treating this as an ordinary quadratic, we get

$$y = \frac{169 \pm \sqrt{(9k^2 - 680)}}{9}$$

Again, the expression under the square-root sign must be a square,
and we have to find values of k that will make it so. Trial of
successive values would be laborious, but there are shorter methods:

Let $9k^2 - 680 = t^2$, where t is any whole number that will satisfy the square condition.

Then $9k^2 - t^2 = 680$, i.e. $(3k + t)(3k - t) = 680$.

Taking the factors of 680, we have:

$680 = 340 \times 2$, or 170×4, or 68×10, or 34×20.

Now tabulate these alternatives, coupled with $(3k + t)$ and $(3k - t)$, where $(3k + t)$ must be greater than $(3k - t)$.

$3k + t =$	340 or	170 or	68 or	34
$3k - t =$	2	4	10	20
so $6k =$	342	174	78	54
$2t =$	338	166	58	14
$k =$	57	29	13	9
$t =$	169	83	29	7
making $y =$	0	28	22	18
whence $x =$		28	23 or 40	23 or 14

But, $x > y$, and $x \neq y$, so $x = 23$.

Hence they sold 23 cars the third week, sales being:

First week: 15 or 11

Second week: 18 or 22

Third week: 23   23'

[Hunter, 1966, problem 6]

568. Naturally, because it will not work. If the specks of mineral were randomly distributed at the start, they will remain randomly distributed when crushed. It is true that particular specks will be closer together, as measured in a vertical direction, as a result of the crushing, but they will also be brought together horizontally, and the distribution will remain random.

Bibliography

Abraham, R. M. (1933) *Diversions and Pastimes*, Constable.

Adams, Morley (1939) *The Morley Adams Puzzle Book*, Faber.

Ainley, S. (1977) *Mathematical Puzzles*, Bell.

Always, Jonathan (1965) *Puzzles to Puzzle You*, Tandem Books.

ApSimon, H. (1984) *Mathematical Byeways in Ayling, Beeling and Ceiling*, Oxford University Press.

Bachet, C. G., de Meziriac (1612) *Problèmes plaisans et délectables qui se font par les nombres*.

Barr, S. (1969) *Second Miscellany of Puzzles*, Dover.

Beasley, J. (1985) *The Ins and Outs of Peg Solitaire*, Oxford University Press.

Beasley, J. (1989) *The Mathematics of Games*, Oxford University Press.

Beiler, A. H. (1964) *Recreations in the Theory of Numbers*, Dover.

Berggren, J. L. (1986) *Episodes in the Mathematics of Medieval Islam*, Springer.

Berlekamp, E., Conway, J. H., and Guy, R. (1982) *Winning Ways*, Vols. 1 and 2, Academic Press.

Black, Max (1952) *Critical Thinking*, Prentice-Hall.

Boyer, C. B. (1985) *A History of Mathematics*, Princeton University Press.

Brandreth, G. (1984) *The Complete Puzzler*, Panther.

Brooke, Maxey (1963) *Coin Games and Puzzles*, Dover.

Budworth, G. (1983) *The Knot Book*, Elliot Right Way Books.

Carroll, Lewis (1958) *Pillow Problems and A Tangled Tale*, Dover

Carroll, Lewis (1961) *The Unknown Lewis Carroll*, ed. S. D. Collingwood, Dover.

Cassell's Book of Indoor Amusements, Card Games and Fireside Fun (1881, facsimile reprint 1973), Cassell.

Cuthwellis, E. (ed.) (1978) *Lewis Carroll's Bedside Book*, Dent.

Delft, P. van, and Bottermans, J. (1978) *Creative Puzzles of the World*, Cassell.

Domoryad, A. P. (1963) *Mathematical Games and Pastimes*, Pergamon.

Dorrie, H. (1965) *100 Great Problems in Elementary Mathematics*, Dover.

Dudeney, H. E. (1907) *The Canterbury Puzzles*, Heinemann.

Dudeney, H. E. (1917) *Amusements in Mathematics*, Nelson.

Dudeney, H. E. (1925) *The World's Best Puzzles*, Daily News, London.

Dudeney, H. E. (1926) *Modern Puzzles*, Pearson.

Dudeney, H. E. (1967) *536 Puzzles and Curious Problems*, Scribner.

Dunn, A. (ed.) (1980) *Mathematical Bafflers*, Dover.

Dunn, A. (ed.) (1983) *The Second Book of Mathematical Bafflers*, Dover.

Eastaway, R. (ed.) (1982) *Enigmas: Puzzles from the New Scientist*, Arlington Books.

Emmett, E. (1976) *The Puffin Book of Brainteasers*, Penguin.

Etten, Henry van (1624) *Mathematical Recreations*.

Eves, H. (1976) *Introduction to the History of Mathematics*, Holt, Rinehart and Winston.

Eves, H. (1981) *Great Moments in Mathematics (Before 1650)*, Mathematical Association of America.

Fauvel, J. and Gray, J. (eds) (1987) *The History of Mathematics: A Reader*, Open University Press.

Fisher, J. (ed.) (1973) *The Magic of Lewis Carroll*, Penguin.

Gamow, G. and Stern, M. (1958) *Puzzle Math*, Macmillan.

Gardner, M. (1966) *More Mathematical Puzzles and Diversions*, Penguin.

Gardner, M. (1971) *Sixth Book of Mathematical Recreations from Scientific American*, W. H. Freeman.

Gardner, M. (1977) *Further Mathematical Puzzles and Diversions*, Penguin.

Gardner, M. (1978a) 'Mathematical Games: Puzzles and Number-Theory Problems Arising from the Curious Fractions of Ancient Egypt', *Scientific American*, October 1978.

Gardner, M. (1978b) *Mathematical Carnival*, Penguin.

Gardner, M. (1981) *Mathematical Circus*, Penguin.

Gardner, M. (1983) *Wheels, Life and Other Mathematical Amusements*, W. H. Freeman.

Gardner, M. (1986) *Knotted Doughnuts and Other Mathematical Amusements*, W. H. Freeman.

Gardner, M. (1989) *Penrose Tiles to Trapdoor Cyphers*, W. H. Freeman.

Gillings, R. J. (1972) *Mathematics in the Time of the Pharaohs*, MIT Press.

Golomb, S. W. (1965) *Polyominoes*, Scribners.

Graham, L. A. (1963) *Ingenious Mathematical Problems and Methods*, Dover.

Graham, L. A. (1968) *The Surprise Attack in Mathematical Problems*, Dover.

Greek Anthology, The (1941) Vol. 5 of the Loeb Classical Library, Heinemann.

Hadfield, J. (ed.) (1939) *The Christmas Companion*, Dent.

Hardy, G. H. (1969) *A Mathematician's Apology*, Cambridge University Press.

Honsberger, R. (1973) *Mathematical Gems*, Mathematical Association of America.

Honsberger, R. (1978) *Mathematical Morsels*, Mathematical Association of America.

Hooper, William (1774) *Rational Recreations*.

Hudson, D. (1954) *Lewis Carroll, An Illustrated Biography*, Constable.

Hunter, J. A. H. (1966) *More Fun with Figures*, Dover.

Hunter, J. A. H. and Madachy, J. S. (1963) *Mathematical Diversions*, Van Nostrand.

Hutton, Charles (1840) *Recreations in Mathematics and Natural Philosophy, translated from Montucla's edition of Ozanam by Charles Hutton.* (New revised edition by Edward Riddle.)

Jackson, John (1821) *Rational Amusements for Winter Evenings.*

Kasner, E. and Newman, J. (1949) *Mathematics and the Imagination*, Bell & Sons.

Kendall, P. M. H. and Thomas, G. M. (1962) *Mathematical Puzzles for the Connoisseur*, Griffin.

Kordemsky, B. A. (1972) *The Moscow Puzzles*, Penguin.

Kraitchik, Maurice (1955) *Mathematical Recreations*, George, Allen & Unwin.

Larsen, L. C. (1983) *Problem Solving through Problems*, Springer.

Lemon, Don (1890) *Everybody's Illustrated Book of Puzzles*, Saxon, London.

Lewis, Reverend Angelo John (Professor Hoffman) (1893) *Puzzles Old and New*, Warne.

Leybourn, Thomas (n.d.) *The Mathematical Questions Proposed in the Ladies' Diary, and their original answers, together with some new solutions, from its commencement in the year 1704 to 1816.*

Li Yan and Du Shiran (1987) *Chinese Mathematics, A Concise History*, trans. J. H. Crossley and A. W-C. Lun, Oxford University Press.

Lindgren, H. and Frederickson, G. (1972) *Recreational Problems in Geometric Dissections and How to Solve Them*, Dover.

Litton's Problematical Recreations (1967) Litton Industries.

Loyd, S. (1959) *Mathematical Puzzles of Sam Loyd*, ed. M. Gardner, Dover.

Loyd, S. (1960) *More Mathematical Puzzles of Sam Loyd*, ed. M. Gardner, Dover.

Loyd (Jnr), S. (1914) *Sam Loyd's Cyclopaedia of 5000 Puzzles, Tricks and Conundrums*, Bigelow, New York.

Loyd (Jnr), S. (1928) *Sam Loyd and his Puzzles*, Barse & Co., New York.

Lucas, Edouard (1883–94) *Récréations Mathématiques*, 4 Vols., Gauthiers-Villars.

Lukacs, C. and Tarjan, E. (1982) *Mathematical Games*, Granada.

Madachy, J. S. (1966) *Mathematics on Vacation*, Scribner.

Mahavira, (1912) *The Ganita-Sara-Sangraha of Mahavira*, trans. M. Rangacarya, Government Press, Madras.

Mauldin, R. D. (ed.) (1981) *The Scottish Book*, Birkhauser.

Mensa (1975) *The Second Mensa Puzzle Book*, Mensa.

Midonick, H. (1965) *The Treasury of Mathematics*, Vol. I, Penguin.

Midonick, H. (1968) *The Treasury of Mathematics*, Vol. 2, Penguin.

Mikami (1964) *History of Mathematics in China and Japan*, Chelsea.

Morris, Ivan (1970) *The Ivan Morris Puzzle Book*, Penguin.

Morris, Ivan (1972) *Foul Play and Other Puzzles*, The Bodley Head.

Mosteller, F. (1987) *Fifty Challenging Problems in Probability*, Dover.

NCTM (1965) *Mathematical Challenges*, National Council of Teachers of Mathematics, USA.

NCTM (1974) *Mathematical Challenges Plus Six*, National Council of Teachers of Mathematics, USA.

NCTM (1978) *Games and Puzzles for Elementary and Middle School Mathematics*, National Council of Teachers of Mathematics, USA.

Needham, J. (1959) *Science and Civilisation in China*, Vol. 3: *Mathematics and the Sciences of the Heaven and the Earth*, Cambridge University Press.

Neugebauer, O. and Sachs, A. (1945) *Mathematical Cuneiform Texts*, American Oriental Society, New Haven.

Newman, D. J. (1982) *A Problem Seminar*, Springer.

Northrop, E. P. (1960) *Riddles in Mathematics*, Penguin.

O'Beirne, T. H. (1965) *Puzzles and Paradoxes*, Oxford University Press.

Ogilvy, S. (1966) *Excursions in Number Theory*, Oxford University Press.

Ore, Oystein (1948) *Number Theory and its History*, McGraw-Hill.

Peet, T. E. (1923) *The Rhind Papyrus*, University of Liverpool Press.

Perelman, Ya. (1979) *Figures for Fun*, Mir Publishing House, Moscow.

Phillips, Hubert (1932) *The Week-End Problems Book*, Nonesuch Press.

Phillips, Hubert (1933) *The Playtime Omnibus*, Faber.

Phillips, Hubert (1934) *The Sphinx Problem Book*, Faber.

Phillips, Hubert (1936) *Brush Up Your Wits*, Dent.

Phillips, Hubert (1937) *Question Time*, Dent.

Phillips, Hubert (1960) *Problems Omnibus*, Arco Publications.

Rouse Ball, W. W. (1974) *Mathematical Recreations and Essays*, 12th edition, ed. Coxeter, University of Toronto Press.

Sandford, Vera (1930) *A Short History of Mathematics*, Houghton Mifflin.

Schuh, F. (1968) *Master Book of Mathematical Recreations*, Dover.

Silverman, D. (1971) *Your Move*, McGraw-Hill.

Sprague, R. (1963) *Recreations in Mathematics*, trans. T. H. O'Beirne, Blackie.

Steinhaus, H. (1963) *100 Problems in Elementary Mathematics*, Pergamon.

Summers, G. J. (1968) *New Puzzles in Logical Deduction*, Dover.

Szurek, M. (1987) *Opowieści Matematyczne*, Wydawnictwa Szkolne i Pedagogiczne, Warsaw.

Thomas, I. (trans.) (1980) *Greek Mathematical Works*, 2 Vols., Heinemann.

Tocquet, R. (1957) *The Magic of Numbers*, A. S. Barnes, New York.

'Tom Tit', (n.d.) *Scientific Amusements*, trans. and adapted by C. G. Knott, Thomas Nelson.

Trigg, C. (1985) *Mathematical Quickies*, Dover.

Tripp, C. (1975) 'Adventitious Angles', *The Mathematical Gazette*, Vol. 59, No. 408, June 1975.

Wells, Carolyn (1918) *Every Child's Mother Goose*.

Wells, D. G. (1975) 'On Gems and Generalisations', *Games and Puzzles*, No. 37, June 1975.

Wells, D. G. (1979) *Recreations in Logic*, Dover.

Wells, D. G. (1983–6) *Can You Solve These?*, Series 1, 2 and 3, Tarquin Publications.

Wells, D. G. (1987) *Hidden Connections, Double Meanings*, Cambridge University Press.

White, Alain C. (1913) *Sam Loyd and His Chess Problems*, Whitehead and Miller, Leeds.

Williams, W. T. and Savage, G. H. (n.d.) *The Strand Problems Book*, Newnes.

Williams, W. T. and Savage, G. H. (1946) *The Third Penguin Problems Book*, Penguin.

Workman, W. P. (1920) *The Tutorial Arithmetic*, University Tutorial Press.

Index

Numbers refer to pages, not to the numbering of the puzzles.